SYSTEMIC

SYSTEMIC

How Racism Is Making Us Sick

By Layal Liverpool

ASTRA HOUSE
New York

Astra House
A Division of Astra Publishing House
astrahouse.com
Printed in the United States of America

Library of Congress Cataloging-in-Publication Data

Names: Liverpool, Layal, author.
Title: Systemic : how racism is making us sick / by Layal Liverpool.
Description: First edition. | New York : Astra House, [2024] | Includes bibliographical references and index. | Summary: "A science-based, data-driven, and global exploration of racial disparities in health care access by virologist, immunologist, and science journalist Layal Liverpool, arguing that racism affects our biology"—Provided by publisher.
Identifiers: LCCN 2023053833 (print) | LCCN 2023053834 (ebook) | ISBN 9781662601675 (hardcover) | ISBN 9781662601682 (epub)
Subjects: LCSH: Discrimination in medical care. | Minorities—Medical care. | Minorities—Medical care—Social aspects. | Health services accessibility—Social aspects. | Social medicine.
Classification: LCC RA563.M56 L58 2024 (print) | LCC RA563.M56 (ebook) | DDC 362.1089—dc23/eng/20231229
LC record available at https://lccn.loc.gov/2023053833
LC ebook record available at https://lccn.loc.gov/2023053834

First edition
10 9 8 7 6 5 4 3 2 1

Design by Alissa Theodor
The text is set in Bembo Std.
The titles are set in ABC Diatype Semi-Mono Trial.

To Mum and Yaz

Contents

Introduction

When I was a teenager, I started to notice small patches of pigment loss on the skin of my face and arms. I was already quite self-conscious, and this made it worse. The patches were accompanied by itching, so my mother took me to see a doctor. Over the next few years, I bounced between health professionals, including dermatology specialists, who prescribed me all sorts of medications, from antibiotics to antifungal treatments. None of it helped. The light patches on my skin not only remained but started to grow larger and more noticeable. In the meantime, I had become an adult and moved from the Netherlands, where I grew up, to the UK. While I was working as a biomedical research assistant in London, I experienced a particularly bad flare-up and was eventually seen by yet another dermatologist. By this point I had seen so many doctors and tried so many treatments that I had grown used to the idea that I must have some extremely rare skin condition that was impossible to diagnose or treat.

After a quick chat and examination of my skin, the new dermatologist told me that I had a classic case of atopic dermatitis—also known as eczema, one of the most common skin conditions. He happened to have darker skin like me, and he told me that eczema sometimes causes hypopigmentation (pigment loss), which looks different on darker skin compared to paler skin. He suggested that the other—mainly White*—doctors I had seen over the years may have failed to recognize

* Throughout this book, I have chosen to capitalize the first letter of the word when referring to racial and ethnic categories: "Black," "Indigenous," "White," etc. There are a couple of reasons for this. One is a practical reason: to distinguish between identities and other descriptors. To give an example, I identify as Black, but I would describe my skin color as brown. Secondly, I have opted for consistent capitalization because I feel that capitalizing all racial and ethnic identities is fair and consistent. I have chosen to write

that I had eczema, because they aren't as used to seeing how it manifests on skin like mine. Thankfully, there are lots of treatments available, and while I still experience flare-ups occasionally, my eczema is largely under control, and my skin pigment has mostly returned.

But the experience stuck with me. It made me wonder whether other medical conditions might be going undiagnosed in darker-skinned people because doctors aren't trained to recognize the diverse ways that their symptoms can manifest. More serious problems, perhaps. Fortunately, I was not the only person who was wondering.

From a young age, Malone Mukwende was interested in science and particularly in the different ways illness can manifest changes in the body. His curiosity motivated him to study medicine. But during his studies at St. George's Hospital, University of London, he noticed a worrying trend in the way medical students were being taught to recognize and diagnose conditions: specifically, an absence of imagery in the study materials showing the way symptoms might appear on darker skin. "It was clear to me that certain symptoms would not present the same on my own skin," Mukwende told the *Washington Post* in an interview in 2020. "I knew that this would be a problem."[1]

Mukwende decided to do something about it. He raised his concerns with lecturers at his university, and together they came up with the idea of assembling a handbook to help doctors. The book, titled *Mind the Gap: A Handbook of Clinical Signs in Black and Brown Skin*, is now freely available to download online. It has been praised by doctors and researchers, including Patricia Louie, a sociologist at the University of Washington whose research in 2018 showed that almost 75 percent of the images found in widely used medical literature in the US showcased lighter skin tones, with fewer than 5 percent featuring darker skin. Even in South Africa, which has a majority Black population, most medical

---

White with a capital *W*, just as in other racial and ethnic identities, because I don't believe that White people exist outside of racialization and racism. The only times I have deviated from this style choice are in instances where I have quoted from written sources, in which case I have reflected the style used by the original sources in question.

students learn from similarly biased textbooks. And there is no doubt that these biases harm people. In the US, for instance, patients from racial and ethnic minority backgrounds are more likely than White people to have common skin conditions, such as psoriasis, go undiagnosed.[2]

The problem of racial bias in medicine is not restricted to dermatology, either. Plenty of studies have found evidence of implicit or unconscious bias among health care professionals more broadly. This isn't surprising—health care workers are human beings, and all of us have biases. But left unchecked, harmful stereotypes, such as the belief that Black people feel less pain than White people, can result in physicians underestimating the severity of symptoms in Black patients and failing to recommend appropriate treatment. Many of these stereotypes stem from the false belief that there are biological differences between races. For instance, a 2016 US study of 222 White medical students and residents found that about half of them endorsed at least one of a list of false statements about biological differences between Black and White people, including "Black people's nerve-endings are less sensitive than White people's nerve-endings" and "Black people's skin has more collagen (i.e., it is thicker) than White people's." The study found that these false beliefs predicted racial biases in the assessment of pain in fictional patients and in subsequent treatment recommendations.[3]

Race isn't a biological characteristic. But I can forgive anyone—including the medical students and residents in the study described above—for looking at our world and concluding otherwise. Across the globe, there are vast disparities between racial and ethnic groups when it comes to health; belonging to a marginalized group is generally associated with poorer health outcomes. I became acutely aware of this as a science journalist reporting on racial and ethnic disparities in infections and deaths that emerged early on during the Covid-19 pandemic in many countries. Some scientists were quick to speculate that these disparities might be at least partly attributable to genetic differences between racial and ethnic groups. This reflects a long-standing trend within medical science of leaning on biology and genetics to try to explain racial and ethnic differences in health outcomes, rather than fully acknowledging and exploring the role of racism itself. Racial bias in medicine represents just one example of the many different ways in which racism is harming our health. Indeed, a growing and undeniable body of evidence points toward racism within our societies as a major contributor to persistent racial and ethnic health gaps in today's world.[4]

Global scientific and medical authorities are increasingly recognizing racism as the dangerous public health threat that it is. In the years leading up to the publication of this book, the renowned medical journal *The Lancet* described racism as a "public health emergency of global concern" and published a special series showing how racism, xenophobia, discrimination, and the structures that support them are detrimental to health. In the same vein, the World Health Organization (WHO) acknowledged in a 2022 report that "health inequities affecting populations experiencing racial discrimination are significant and remain pervasive across many countries," noting that these inequities are "driven by interactions between the wider social determinants of health and structural racism and discrimination in society."[5]

But mainstream conversations about race and health are lagging. In a 2020 interview with NPR, Louisiana senator Bill Cassidy attributed the disproportionate impact of Covid-19 on the state's Black population to the fact that "African Americans are 60% more likely to have diabetes," without acknowledging how racism contributes to that underlying health disparity. Responding to the suggestion that such health disparities were rooted in years of systemic racism, Cassidy, who is also a doctor, said: "Well, you know, that's rhetoric, and it may be. But as a physician, I'm looking at science." Meanwhile in the UK, a 2021 report commissioned by the government suggested that research into ethnic health disparities should consider "genetic and biological differences" and dismissed the role of racism as a key driver of health inequalities. The lag in mainstream discussions about race and health is hindering vital efforts to research and address racial and ethnic health gaps in the US, the UK, and globally. I want to help change that.[6]

This book is an investigation of the heavy toll that racism takes on health, split into four sections. The first will map out the vast health gaps that exist between racial and ethnic groups within different countries across the globe. Across two chapters we will expose a worrying pattern of people belonging to marginalized groups tending to experience poorer health across the board—starting from pregnancy, childbirth, and early life and, for those who survive, continuing across the life course with persistent racial and ethnic disparities in terms of who becomes ill and who dies as a result. The remaining three sections of the book will focus on *why* these health inequities exist and persist in the first place.

We will examine the multiple, overlapping ways in which racism harms health, starting in the second section of the book with a close look at the insidious public health threat posed by systemic and interpersonal racism in daily life. We will learn how the systems, structures, and institutions that underpin societies globally mean that not everyone has equal access to a healthy living environment, and we will discover the devastating toll that chronic racism-related stress and acute racial trauma can take on health and well-being.

In the third section, we will continue our exploration of how racism in medicine—including racial bias among health care workers and racism embedded into medical education, practices, and guidelines—reinforces existing racial and ethnic health inequities.

Finally, in the fourth section, we will turn our attention to the scientific research that underpins our understanding of our health, including health inequities. We will dig into the data—and glaring data gaps—in health and medical research, and we will discover why bias in science is bad for everyone's health.

I said *everyone* on purpose because this problem affects us *all*, whether we realize it or not. Racism as it manifests across societies globally is not only making us collectively more ill—it is also holding back our medical research efforts, reducing the potential for scientific discoveries that could help us to be treated more effectively when we do become unwell.

I want to take this opportunity to note that although this book chiefly concerns itself with the health harms of racism, all forms of discrimination—including on the basis of class, gender identity, sexuality, and disability, to name a few—are harmful to health. Entire books can and have been written about the harms of these various other forms of oppression, and while I will not attempt to recapitulate them in this book, we will occasionally come across instances where racism overlaps or intersects with other types of discrimination in the context of health. This is inevitable: racism is bad for our health and its harms can be particularly acute for people with multiple, intersecting marginalized identities.

I have often thought that if racism were a virus, we would be up in arms at the amount of needless pain, suffering, and death being caused by it worldwide. We would be rushing to develop vaccines and searching for treatments. Racism isn't a virus, although it does make human populations more susceptible to them—something we observed recently when Covid-19 spread across the globe. Pockets of inequality within societies, including those driven and perpetuated by racism, provide cracks through which infectious disease epidemics can creep. These cracks are present all the time but become particularly pronounced during health crises such as pandemics. In fact, racism is an especially insidious threat to our health precisely because it is so easily overlooked or, in some cases, perhaps willfully ignored. Yet, as we will discover, racism is embedded deep into the foundations, not only of our societies, but of medicine and science—hidden in plain sight. And, in a painful irony, it is something that we humans inflicted upon ourselves.

Racism is a belief system that was invented and spread around the world during an era of European imperialism and expansion. This is not a coincidence: designating groups of people as biologically different from and even inferior to others helped to justify their oppression, through colonization and transatlantic slavery. Several brilliant books—including Dorothy Roberts's *Fatal Invention*, Angela Saini's *Superior*, Adam Rutherford's *How to Argue with a Racist*, and Annabel Sowemimo's

Divided—dive deeply into the inventions of race and racism within science, but I still think this influential chapter in scientific and human history warrants a brief summary here.[7]

Some of the earliest scientific attempts to categorize people into racial groups were made in the seventeenth and eighteenth centuries. In a thesis published in 1795, German anthropologist Johann Blumenbach divided humans into five ancestral groups: Caucasians, Mongolians, Ethiopians, Americans, and Malays. With minor variations, these categories would go on to be used by scientists to perpetuate racism and promote eugenics for centuries to come. Charles Darwin's half cousin, the eugenicist Francis Galton, drew on them in the nineteenth and twentieth centuries, with catastrophic consequences. And similar practices of grouping people—and ourselves—into racialized categories persist globally to this day.[8]

Interestingly, Blumenbach himself refuted the idea that Africans (whom he referred to as Ethiopians) were inferior to White Europeans (whom he called Caucasians), putting him somewhat at odds with the prevailing viewpoint of his time. He did, however, write that the Caucasian skull of a Georgian female was the "most handsome and becoming" of those which he analyzed during his research. He also theorized—incorrectly—that white, as a skin color, was "the primitive color of mankind," arguing that changing from lighter to darker skin would have been easier than the reverse.[9]

We now know this to be untrue. The current scientific consensus is that *Homo sapiens* originated in Africa around three hundred thousand years ago and then migrated outward—all modern humans are descended from African *Homo sapiens* populations. Pale skin is thought to have become more common after some humans migrated out of Africa, possibly as an adaptation to reduced sun exposure in northern climates. There is nothing particularly unique or special about this. The genes that influence skin color—or those that influence other traits typically associated with race, such as nose and eye shape or hair texture—are not any more biologically significant than the thousands of other protein-coding genes that influence the rest of human biology. Crucially, more genetic variation exists within human populations from the same large geographic regions than exists between these populations.[10]

In 1972, evolutionary biologist and geneticist Richard Lewontin illustrated the scientific fallacy of biological race in a seminal paper titled "The Apportionment of Human Diversity." When the news broke that Lewontin had passed away in July 2021, aged ninety-two, I was working as a journalist writing for the science magazine *New Scientist* and happened to be deeply engrossed in an investigation on scientific and medical racism. One of the many tabs I had open on my computer was a PDF of his paper, in which he unambiguously demonstrates that far more genetic variation exists among human beings within the commonly defined racial groups than between these groups.[11]

Lewontin describes these broad racial groups as "African peoples, European nationalities, Oceanian populations, Asian peoples, and American Indian tribes" and he makes a point of putting the term "races" in quotation marks, presumably to emphasize the fact that his work completely undermines the use of these categorizations within science in the first place. Specifically, Lewontin analyzed variation in blood proteins—as a marker of variation in the genes that encode them—across human populations from around the world. He found that almost 94 percent of human genetic variation exists within populations or within so-called racial groups, compared to just over 6 percent present between groups. Lewontin's findings were extremely powerful, because they put numbers on the growing scientific understanding of human genetic variation.[12]

Another landmark study that confirmed the invalidity of biological race was published by population geneticist Noah Rosenberg, now at Stanford University, and colleagues in 2002. Rosenberg and his team analyzed DNA from more than 1,000 people across 52 geographic regions around the world. They found that genetic differences between individuals within the same geographic group accounted for 93 to

95 percent of all genetic variation, while differences among major geographic groups constituted just 3 to 5 percent.[13]

Rosenberg and his colleagues then plugged their data into a computer program called STRUCTURE and told it to break down the results into defined numbers of clusters of DNA. Instructing it to divide the entire human population into five clusters led it to group people into Africans, Europeans and Middle Easterners, eastern Asians, Americans and Australians—groupings not too dissimilar from Blumenbach's or from the racial categorizations commonly used today. In fact, the study was reported in some media outlets at the time as having demonstrated the biological reality of race. But as US law and sociology scholar Dorothy Roberts points out in her book *Fatal Invention*, the number of genetic clusters was completely arbitrary: increasing the number of clusters beyond five resulted in the computer putting smaller, more isolated populations into separate clusters, with the degree of overlap between clusters becoming increasingly evident. Sampling people from more geographic regions would probably emphasize this blending between clusters even further. In his book *A Brief History of Everyone Who Ever Lived*, geneticist Adam Rutherford explains that this research ultimately demonstrates that the question of how many races there are is "meaningless." Genetics tells us it is much more accurate to view humanity on a spectrum than to attempt to manufacture categories of people with hard boundaries between them.[14]

It is for this reason that attempts to use biology or genetics to explain racial and ethnic disparities in health are so problematic. Jay Kaufman, an epidemiologist at McGill University in Canada, is all too aware of this. Kaufman has done extensive research on racial and ethnic disparities in health and in medical treatment, including in Canada and the US. He explained to me the problem with conflating socially defined categories, such as race, with biological ones, like genetics. "We make categorizations of race based on continental populations," Kaufman told me. "So, the definition in the US census for Black people is people from sub-Saharan Africa. And, obviously, that's a huge number of people—almost a billion people. And it's people with very, very different local adaptations, very, very different physique, and diet, and social history, and biological history and so forth. They are no more closely related than Africans to Europeans, on average," he explained. "There's just not any coherent, physiologic characteristic of all White people, or of all Black people, aside from skin color," he said.

Even skin color fails to consistently group people into the same racial or ethnic categories around the world. I identify as both Black and Mixed, but I often joke that while I have always been perceived as Black in the various European countries in which I have lived, when I visit Ghana people often call me "Oburoni," a Twi word used colloquially to mean "White person." My skin color—which is brown, by the way—obviously doesn't change drastically depending on where in the world I am. Neither does my biology or genetics. But social categorizations like race *do* change—with both time and place. The South African comedian and former talk show host Trevor Noah has joked about his experience of being "upgraded" from "colored," or Mixed, to "Black" when he moved from South Africa to the US. Indeed, while apartheid-era South Africa classified Mixed people as "colored," in the Jim Crow US the "one-drop rule" asserted that any person with even one Black ancestor should be legally considered Black. This is a great example of why racial categories are illogical and inconsistent ways of grouping people from a scientific or medical standpoint.[15]

Racism isn't only illogical; it is also one of the least widely acknowledged and most dangerous public health threats facing our world today—and it is high time that all of us recognize and name it as such. We should listen to the many people who have been sounding the alarm on this for decades, a number of whom we will get to know in this book. These people—many of them belonging to too-often-marginalized racial or ethnic groups—include not only doctors and academic researchers such as Kaufman, but also medical students and patients, campaigners and activists, innovators and policymakers.

With their help, I have delved into the vast amount of research and data that illustrates just how damaging racism is to health, and—by sharing the experiences of some of those people who are worst impacted—I will explain here how that damage manifests itself in the human body and mind. I will show that health disparities, along lines of race and ethnicity, can be found all around the world—but also that they are by no means inevitable. Plenty of people globally are working to tackle racism in all its forms and eliminate health inequities, including many people within the medical and scientific establishments. We can support them in their efforts, join them in their calls for justice, and collectively pivot toward a fairer and healthier future.

PART I

Health Gaps

CHAPTER 1

Pregnancy and Childbirth

Serena Williams is one of the most successful people in the world. You don't have to take my word for it: the twenty-three-time Grand Slam winner regularly appears on annual lists of the world's most influential people published by the likes of *Time* and *Forbes*. But no amount of success, wealth, or fame could protect Serena from what happened to her in September 2017.

A day after the birth of her daughter, Alexis Olympia, Serena—who has a history of blood clots in the lungs, called pulmonary embolisms—suddenly began feeling short of breath. Serena had ended up delivering her daughter by emergency cesarean section, and due to the surgery, she had come off her routine regimen of blood-thinning medication. Because of this, and given her history of blood clots, she immediately assumed that her symptoms the following day were the first signs of a pulmonary embolism. "Serena lives in fear of blood clots," Rob Haskell wrote in a 2018 profile of the athlete published in *Vogue*. She stepped outside her room in the hospital so as not to worry her mother and informed the nearest nurse that she needed a CT scan with contrast, a substance that helps visualize blood vessels, and intravenous heparin, a blood thinner, as soon as possible—all the while gasping for air. The nurse dismissed her at first, thinking that the pain medicine she had taken might have left her a bit dazed. But Serena was insistent. Eventually, she was seen by a doctor who began performing an ultrasound of her legs. "I was like, a Doppler [ultrasound]? I told you, I need a CT scan and a heparin drip," Serena recalled telling the team at the hospital. After the ultrasound failed to reveal anything, she was finally sent for the CT scan. The scan revealed what she had feared all along—several small blood clots had settled in her lungs. Serena was put on the drip within minutes. "I was like, listen to Dr. Williams!" she told Haskell.[16]

Unfortunately, Serena's problems didn't end there. Intense coughing spells induced by the blood clots in her lungs caused her C-section wound to tear open, and during a subsequent surgery, doctors discovered that a serious injury called a hematoma had filled her abdomen as a side effect of the blood thinner. She had to undergo yet another surgery to have a filter inserted into a major vein, to prevent further clots from entering her lungs. Serena was discharged home a week later and spent the next six weeks—her newborn child's first weeks in the world—unable to get out of bed.

This is a single story, of course, and not evidence on its own. But it made an impact. Serena's experience felt all too familiar to me and to many other Black women and birthing people*, some of whom shared their personal experiences of being dismissed, doubted, or disbelieved by doctors and nurses in an outpouring of responses on social media at the time. Serena used her platform to highlight the fact that Black women in the US are more likely than White women to die from complications in pregnancy or childbirth. The evidence is harrowing: according to the US Centers for Disease Control and Prevention (CDC), non-Hispanic Black women died three times as often from pregnancy or childbirth complications compared with non-Hispanic White women between 2014 and 2017 (in absolute terms, 42 Black women died for every 100,000 live births, compared with 13 White women). Over the

---

* I have used the phrase "women and birthing people" here in recognition of the fact that people who don't identify as women may become pregnant and give birth. However, this chapter refers to both "women and birthing people"' and "women." That is because, in cases where I am referencing sources, such as scientific research articles, I have reflected the language used by the original source.

same period, non-Hispanic American Indian or Alaska Native women died at twice the rate of White women. Serena described the statistics as "heartbreaking" during a BBC interview in 2018. "Doctors aren't listening to us," she said.[17]

Galvanized by Serena's story and the response it received online, I decided to dive deeper into the numbers. Numerical statistics can be numbing, but they also form an essential bedrock for understanding the scale of the problem. I wanted to map out the shocking health gaps that start to appear from the very beginning of life—and even before life has begun. From pregnancy to childbirth to early life, racial and ethnic health gaps exist in countries all around the world. We will dig into the data on this and meet some of the real people behind the numbers—and we'll hear from researchers who are investigating how racism contributes to these devastating inequities.

In the US, the stark statistics that Serena's story drew attention to are from the CDC's Pregnancy Mortality Surveillance System, which defines a pregnancy-related death as "the death of a woman while pregnant or within 1 year of the end of pregnancy from any cause related to or aggravated by the pregnancy." The CDC has been actively collecting these data since 1987 by analyzing death records and linked birth records. According to the CDC, it wasn't possible to reliably separate the statistics into racial-ethnic categories beyond "Black," "White," and "Other" prior to 2006, because there weren't sufficient data for other racial and ethnic groups. This changed over time as more data became available, but one thing has remained worryingly persistent: the gap in the pregnancy-related death rate between Black and White women. Data from 1987 all the way through to 2017 show that the pregnancy-related death rate has consistently remained between about three and four times higher among Black women compared with White women.[18]

After 2017, things don't look much better. The CDC's National Vital Statistics System also collects data on maternal mortality in the US, in this case defined as "the death of a woman while pregnant or within 42 days of termination of pregnancy," excluding deaths from accidental or incidental causes. Despite this slightly more restrictive definition, the pattern in the data remains. In 2019, the most recent year for which data were available at the time of writing, non-Hispanic Black women died more than twice as often during pregnancy or childbirth compared with non-Hispanic White women. Separate research suggests the risk of severe maternal morbidity—life-threatening complications during pregnancy—and mortality is about twice as high among Indigenous women in the US compared with non-Hispanic White women.[19]

It was only a couple of years later, however, when I spoke about Serena Williams's ordeal with Michele Evans, an oncologist who researches racial and ethnic health disparities at the National Institutes of Health in Maryland, that I finally recognized why it was particularly striking that something like that could have happened to the tennis champion. "Look at Serena Williams, very wealthy, an athlete," said Evans. Serena's experience suggests the idea that you can buy your way out of racism is mistaken. "It's not true," Evans told me. "It's not just low SES [socioeconomic status] women. It's high SES women" who are affected, she said. This is reflected in the data: there is evidence that when it comes to complications and mortality in pregnancy and childbirth, there are racial and ethnic disparities even in the highest income and educational groups.

A 2019 study found that among women with a graduate-level education or higher, the pregnancy-related death rate for Black women was still more than five times higher compared with White women. In fact, the rate of pregnancy-related death among Black women who were university graduates or above was almost two times higher than that among White women educated to high-school level or below. This is not to say that educational level—which is strongly associated with wealth and income—is irrelevant. To the contrary: socioeconomic status, whether measured by educational level or income, is clearly associated with overall health. Across all racial and ethnic groups, pregnancy-related death rates did generally fall with increasing educational level from high school upward. But—critically—this protective effect of socioeconomic status was stronger for White women compared with Black or Indigenous women, suggesting additional factors are at play.[20]

Serena was acutely aware of her relative privilege. She told the BBC that her ordeal would have been even more difficult without the relatively good health care she can afford: "To imagine all the other women that do go through that without the same health care, without the same response, it's upsetting."

She was right about the importance of access to health care. According to a 2020 report by March of Dimes, a nonprofit organization, about seven million women of childbearing age in the US live in areas where there is no or limited access to maternity care. Further, increasing closures of maternity wards over the past decade or so have disproportionately disadvantaged Black women. A 2017 study documented this decline in hospital-based obstetric services in rural counties across the US between 2004 and 2014. It found that counties with higher proportions of non-Hispanic Black women and lower median household incomes had higher odds of losing all hospital obstetric services. J'Mag Karbeah, a health services researcher at the University of Minnesota, has long been concerned about this pattern. "How can you say you care about birth outcomes and Black people's birth outcomes, if you keep closing maternity units in Black neighborhoods?" she said. People are less likely to be able to attend prenatal appointments if they have to travel an hour or more to access the care that they need, she pointed out.[21]

In addition to access, the quality of health care matters. Research in the US suggests that Black women are more likely than White women to deliver their babies in hospitals with high rates of severe maternal morbidity and that perform poorly on indicators of obstetric care quality and patient safety. A 2015 study of data from more than five thousand hospitals across the US found that one-quarter of the hospitals provided care for three-quarters of all the deliveries by Black women between 2010 and 2011. These so-called "high Black-serving" hospitals had higher severe maternal morbidity rates compared with other hospitals. This is no coincidence: these hospitals are often in neighborhoods disadvantaged by historical racial segregation policies. Karbeah pointed out that in Minnesota, where she is based, racial covenants preventing non-White people from buying homes in certain neighborhoods were only made illegal in 1962. When looking at racial and ethnic health disparities today, we mustn't forget the impacts that history has had and continues to have, she said.[22]

Karbeah explained that attempts to try to attribute racial and ethnic disparities in pregnancy-related mortality solely to class or socioeconomic status are missing the point. "There's a lot of folks who love to say, 'It's not race, it's class,' right, that's driving all of this," she said. "But we know that it's not, point blank, it's not."

As we will discover in chapter 3, systemic racism is a key driver of socioeconomic inequality in the US and elsewhere. Instead of simply stating that Black people are dying disproportionately because they are poor, we should be asking *why* Black people in the US are disproportionately poor in the first place. It is impossible to answer that question without reckoning with the country's racist history and continuing legacy of systemic racism. This is the lens through which we need to look at racial and ethnic gaps in maternal health, said Karbeah. "It's a biological consequence of living in a racist environment."

After the media hype around Serena Williams's story died down, I found myself unable to stop thinking about her experience and how many people had related to it—as well as those grim statistics on racial and ethnic disparities in pregnancy-related mortality in the US. I wondered whether similar data were available for the UK, where I was working on my PhD at the time. It didn't take long to establish that similar disparities exist in the UK, even though the UK's overall maternal mortality rate is almost four times lower than that in the US.

A study of pregnancy outcomes in the UK between 2017 and 2019 found a more than four-fold higher rate of maternal mortality among women from Black ethnic backgrounds, and an approximately two-fold higher rate among Mixed ethnicity women and among women from Asian ethnic backgrounds, compared with White women. In absolute terms, it found that 32 Black women died among every 100,000 giving birth, compared with 15 Mixed women, 12 Asian women, and 7 White women. Of the disparity between Black and White women, a 2020 UK parliamentary report titled *Black People, Racism and Human Rights* said: "The NHS [National Health Service] acknowledge and regret this disparity but have no target to end it. This must be rectified."[23]

These disparities aren't new. A study that analyzed trends in maternal mortality rates over time from 2009 to 2017 identified similar disparities between Asian and White and between Black and White ethnic groups in the UK, and found that the maternal mortality gap between Black and White women actually widened over time. The Covid-19 pandemic didn't help matters. A study conducted between March 1 and April 14, 2020—early on during the UK's epidemic—found that 56 percent of pregnant women admitted to the hospital with coronavirus in the UK were from a Black, Asian, or other ethnic minority background.[24]

On April 19, 2021, the UK parliament debated a petition calling for maternal mortality rates and health care for Black women in the UK to be improved. The petition was put forward by Tinuke Awe and Clotilde Abe, cofounders of Five X More, a grassroots organization committed to improving Black women's and birthing people's maternal health outcomes in the UK. Bell Ribeiro-Addy, Labour MP for Streatham, shared her own heartbreaking pregnancy experience during the debate.[25]

"During my own pregnancy, it was not hard to find instances where, as a Black woman, how I was perceived or believed drastically impacted the care I received, from complaints about how I was feeling to being denied scans. We know that Black women are perceived to experience less pain. We know this, and we have no target to end it," Ribeiro-Addy told Parliament. "Things went from bad to worse for me. I was swollen. My blood pressure would get so high that I would feel dizzy and my nose would bleed," she recalled. On one occasion, she was feeling so unwell that her doctor had her rushed to the hospital for further tests and scans. On arrival, she was immediately admitted with preeclampsia—a pregnancy complication characterized by high blood pressure. "My last conversation with the consultants was harrowing," she remembered. "They said that my pregnancy had become very dangerous and there were only two outcomes: my child would die, or both myself and my child would die. My diagnosis was too late for any intervention, and simple steps—which I soon found were simple things such as taking aspirin—were no longer an option for me."

The consultants advised a late termination and a delivery to save Ribeiro-Addy's life. They told her that her condition was deteriorating so rapidly that she would immediately have to nominate someone to make the decision on her behalf, should she become unconscious. "I did not have to make this decision, because a scan scheduled the day after that meeting showed that my baby's heart had stopped beating." She was induced and after about eighteen hours of labor her daughter was born. "As a person of faith, even then, I still had faith that maybe the doctors were wrong and everything would be okay, but she did not move, she did not cry, and there was no miracle."

Ribeiro-Addy went on to highlight research which found that between 2005 and 2013, the odds of severe maternal morbidity were 83 percent higher among women of Black African origin—like herself—and 80 percent higher among Black Caribbean women compared with White women in the UK. And that is after adjusting for other factors known to be associated with severe maternal morbidity, including preexisting medical conditions, body mass index (BMI), socioeconomic status, and behavioral factors such as smoking. "Not only do we not have a target to end this," said Ribeiro-Addy. "But we do not have information about the health issues that Black women go on to face."[26]

The US and UK aren't exceptional in this. Disparities in maternal health between racial, ethnic, or caste groups have been recorded in many other countries for which such data are available, including India, China, Brazil, Nigeria, Mexico, and Australia to name a handful. And the problem is compounded by the vast differences in maternal mortality rates between regions of the world. For instance, almost 95 percent of all maternal deaths in 2020 occurred in low- and lower-middle-income countries, according to the World Health Organization. As a result, people of color are disproportionately affected on a global scale too; sub-Saharan Africa and southern Asia accounted for approximately 87 percent of the estimated global maternal deaths in the same year.[27]

D'lissa Parkes was delighted when she learned that she was pregnant. A child practitioner based in London, she loved children and was always the first to volunteer to babysit her nieces and nephews. "She lived for family get togethers such as birthdays and Christmas time where she always looked forward to cosying down on Christmas night with a nice brand new pair of pjs," her mother, Sylvia, wrote in a blog post on the Five X More website. "She was the life and soul of every family time we enjoyed together."[28]

About seven weeks into her pregnancy, D'lissa experienced bleeding and was sent for a scan, which showed a mass in her abdomen. She had a history of severe constipation and had previously been diagnosed with megacolon, a widening of the large intestine. Sylvia said the mass was visible in subsequent scans during D'lissa's pregnancy but wasn't acted upon or linked with her maternity notes. At thirty-seven weeks, the baby was found to be breech. Usually during a first pregnancy—as D'lissa's was—the baby moves into a head-down position and engages with the pelvis naturally at around thirty-seven weeks, ready to be born headfirst. But sometimes this doesn't happen and the baby is in breech position until labor starts. D'lissa went to the hospital for an external cephalic version or ECV—that's where a health care worker tries to turn the baby into a head-down position by applying pressure on the abdomen. After a week, D'lissa returned to the hospital and was seen by the same doctor who had performed the ECV. She was in a lot of pain, Sylvia later recalled in an interview for a BBC documentary, and was one centimeter dilated. At this point, Sylvia said, the baby's head still hadn't engaged, but the doctor insisted that this was normal because Black women's pelvises were shaped differently compared with White women's. D'lissa was sent home.[29]

In the early hours of the following morning, D'lissa went fully into labor and was taken back to the hospital where she was rushed for an emergency cesarean. She died after a blood clot formed in her lungs and caused her heart to stop beating. Her daughter, named S'riaah Christina, survived with a severe brain injury and has been diagnosed with cerebral palsy.

An inquest found that the main risk factor in D'lissa's death was pregnancy. However, the coroner, Andrew Walker, noted errors in her care, including the failure of medical staff to access her medical records that contained information about D'lissa's constipation and her previous diagnosis. D'lissa had a large fecal mass in her large intestine, which had forced her uterus out of shape, but the maternity staff who cared for her weren't aware of this. "This possibly could have contributed minimally to her pulmonary embolism [blood clot in the lungs], but the principal risk factor was pregnancy," Walker found in his verdict. He also said that D'lissa should have been kept in the hospital following her visit the day before her death because there was significant risk that the baby might suffer, but he added that sending her home made no difference to her chances of survival. In her blog post, Sylvia wrote: "I will always maintain that her death could have been prevented if she was taken seriously when she attended the hospital on the day before she died and was sent home. I'm not a medical expert . . . but I'm her mum."[30]

I was devastated when I first learned about what happened to D'lissa. I was twenty-six at the time—coincidentally the same age she was when she died. The Five X More team had shared Sylvia's blog post about losing her daughter on their website on what would have been D'lissa's thirty-first birthday, during the first UK Black women's maternal health awareness week, which they organized in September 2020. The scientist in me kept churning over the medical details, which Sylvia had so generously made public. It was in examining these details more closely that I discovered that the idea that Black women's pelvises are shaped differently to White women's—which D'lissa's doctor referred to when he attended to her just a day before her death—is problematic to say the least.

In 1933, New York–based obstetricians William Caldwell and Howard Moloy published a description of female pelvic anatomy whose legacy persists in modern medical and midwifery textbooks. Caldwell and Moloy analyzed human skeletons from a collection assembled in the early 1900s and categorized female pelvic shape into four subtypes: "gynecoid," "platypelloid," "android," and "anthropoid." These categorizations built upon earlier designations made by the British anatomist William Turner in the 1800s. Turner's work focused on the prevalence of different pelvic shapes in the various "races" as well as their supposed suitability for childbirth. According to Turner, the gynecoid pelvic shape was "characteristic of the more civilized and advanced races of mankind"—namely White Europeans, such as himself—while the anthropoid shape "was more frequently observed in the lower races of man and presented a degraded or animalized arrangement."[31]

The Caldwell-Moloy classification helped transfer these racialized categorizations into the obstetrical literature in the twentieth century, just as hospital birth was becoming more common in the Western world. In 1949, the anthropologist Adolph Schultz at the Johns Hopkins University School of Medicine in Maryland published a paper comparing measurements of pelvic bones between the "adult female Negro" and adult female nonhuman apes and monkeys. Numerous studies in the US also continued to compare pelvic shape between Black and White people. A 2002 study that analyzed eighty skeletons taken from the Hamann-Todd human osteological collection—the same collection used by Caldwell and Moloy in the 1930s—concluded that the pelvic floor area was 5 percent smaller in African American women compared with European American women on average. But at the time those skeletons were collected, specimen collectors categorized them as "Black" or "White" in part based on racialized assumptions about pelvic anatomy. One of the collectors, Thomas Wingate Todd, even noted that "the narrow pelvis is so distinctively a Negro character and our average is so much less than those of other American Negro samples that it may well serve as an indication of relatively pure Negro material." This introduces a clear bias, as in order to identify supposed anatomical differences between racial groups, the collectors first used those very same anatomical differences to confirm race. The authors of the 2002 study note this in their paper as a limitation, writing: "These classifications do not give a complete description of the ancestry of these samples. Race is an abstract term that is restricted to the social interpretations of it."[32]

A more recent analysis of pelvic CT scans from sixty-four female volunteers in Australia found no obvious clustering into the four pelvic types in the Caldwell-Moloy classification, instead reporting "an amorphous, cloudy continuum of shape variation." That study, published in 2015, recommended that teachers and authors of midwifery, obstetrics, and gynecological texts be more cautious about continuing to promote the Caldwell-Moloy classification. Subsequent research has demonstrated huge variation in the shape of the birth canal both within and between human populations in different regions of the world. For instance, a 2018 study that looked at birth canal shape in populations across five continents noted that "canal variation is continuous and without abrupt differences between regions."[33]

Human pelvises come in all shapes and sizes, and that has nothing to do with race. D'lissa's doctor was mistaken—and, unfortunately, he isn't alone. The persistent idea that Black women's pelvises are uniquely different from and even inferior to White women's is just one example of how scientific racism continues to influence medical practice (something we will delve into in more detail later in this book). This particular flavor of scientific racism hasn't only targeted Black women. In nineteenth-century Mexico, obstetrician Francisco Flores y Troncoso described the pelvises of Indigenous women as "downward and backward" and inferior to those of White Spanish women. Flores y

Troncoso even proposed that miscegenation—mixing between races—was the root cause of "downward and backward" pelvises among Mexican women. These ideas were later used to justify propositions of eugenic sterilization due to supposed "pelvic deficiency" among Indigenous women in Mexico.[34]

Racist ideas about pelvic anatomy are part of a wider history of racism and discrimination in obstetrics and gynecology. William Potts Dewees, one of the early pioneers of this field of medicine, believed that wealthy White women suffered more during childbirth compared with poorer White women or Black women, because civilization made their brains more sensitive. Meanwhile, James Marion Sims, who is often referred to as the father of gynecology, performed experimental surgeries on enslaved Black women without using anesthetic. This failure to empathize with or even acknowledge Black women's pain remains a problem today. "It's this commitment to ignoring the humanity of Black people that has us where we are," observed Karbeah. "People simply do not believe [Black women and birthing people] when they say things like, 'I don't feel well,'" she said. "As was the case with Serena Williams, they will not be listened to if they voice concerns."[35]

Medicine's longtime obsession with searching for biological differences between races may be detracting from efforts to investigate the root causes of racial and ethnic disparities in maternal health, according to Jenny Douglas, a public health researcher at the Open University in the UK. When it comes to investigating the factors that contribute toward racial and ethnic disparities in pregnancy and childbirth outcomes, she told me, "We are wasting money looking for biological explanations." In the UK, for example, where Douglas has been researching Black women's health for decades, she said she would like to see a greater emphasis on studies that center Black women's experiences.

"What we should be doing is some more qualitative research with Black women about their experiences of health care and maternity care," she explained. "Black women are not listened to, Black women are not believed, Black women are not cared for in the same way that White women are." Indeed, despite the large disparity in maternal mortality between Black and White women, there has been limited research systematically investigating Black women and birthing people's perceptions of health and maternity care in the UK. "We have put in research proposals, which have not been funded," said Douglas. The grassroots organization Five X More is working toward filling this information gap through the Black Maternity Experiences Survey, launched in April 2021 and led by a team of Black women—including Douglas and other maternal health experts in the UK.

Five X More published the results of its first Black Maternity Experiences Survey in May 2022. The results showed that 43 percent of Black or Black Mixed women reported feeling discriminated against during their maternity care, with a similar proportion rating the care they received during childbirth as "poor" or "very poor." As part of the extensive survey, which included more than one thousand Black or Black Mixed women who had been pregnant and had accessed maternity services in the UK between 2016 and 2021, participants were also able to share personal anecdotes. A common theme that emerged from these stories was that many participants felt dismissed by health care professionals. For instance, a lot of survey respondents reported that their concerns were ignored, leaving them feeling "scared" and as if their "thoughts and feelings do not matter." In some cases, this led to emergency situations, as one participant shared: "I ended up having to go into surgery because my daughter's heartbeat had started to drop [. . .] I feel like we only got to this stage because the first midwife had dismissed me for hours when I was trying to [. . .] tell her that something is wrong."[36]

According to Five X More's report on the survey, many women also described being faced with situations where it was insinuated that they should be able to manage their pain. Other racist assumptions cropped up prior to giving birth too—one woman shared the following anecdote about a health visit in early pregnancy: "First visit a nurse said she was shocked I knew who the father was. As people like me usually don't know." Another shared this of her labor experience: "One midwife when doing the sweep said that the reason for dilation taking so long for me was 'probably due to an African pelvis' [. . .] I was mortified that she actually believed there was such thing as an African pelvis." Five X More's report concluded that the overall maternity care experience for Black women in the UK "is one of racial inequality."

In the same week that Five X More published its report in 2022, the charity Birthrights published its own report based on a yearlong investigation into racial injustice and human rights in maternity care in the UK. It concluded that Black and Asian ethnicity women and birthing people are being harmed by racial discrimination in maternity care.[37]

Earlier research in England has also suggested that ethnic minority women generally have poorer experiences with maternity services compared with White women. A 2010 survey of more than twenty-four thousand women in England found that throughout their maternity care, women from ethnic minority groups were less likely than White women to report feeling spoken to so they could understand, being treated with kindness, being sufficiently involved in decisions, and having confidence and trust in the staff. These findings were echoed in the individual responses submitted anonymously by women who participated in a smaller, open-response survey conducted in England around the same time. In response to that survey, an anonymous Black woman shared: "In the postnatal ward I was ignored and felt that I did not belong to that environment. Therefore, when I was asked to go home after not having enough rest I still went home." And here is another short account shared by an anonymous Asian woman via the same survey: "The midwife did not really appreciate the amount of pain I was in and dismissed this even though I repeatedly said that it was severe and that I and my husband were medical doctors . . . I was sent home inappropriately even though I was in severe pain."[38]

Anecdotes such as these are reflected in statistics from the US, where there has been more research documenting experiences of implicit bias and racism in maternity care and, in some cases, even linking these to poorer birth outcomes, particularly among Black and Indigenous women. Across the country, Black, Indigenous, Hispanic, and Asian women are all more likely than White women to report experiencing at least one form of mistreatment by their health care providers during pregnancy or childbirth. And, according to a survey of women who gave birth in California in 2016, more than 10 percent of Black women report perceived unfair treatment from maternity care providers because of their race, compared with 1 percent of White women. Further, Black women are also more likely than White women to report feeling pressured to have medical interventions during labor and delivery, and to report feeling that hospital staff didn't encourage them to make their own decisions about the birth process. Negative experiences such as these, combined with stereotypes about Black women being angry or difficult, may discourage Black women and birthing people from asking for help when they need it. "If you speak out as a Black woman, you are seen as a problem," said Douglas. "And so, often, Black women may not themselves articulate how they're feeling, because they know that nobody's really listening to them."[39]

In addition to implicit bias and experiences of interpersonal racism in health and maternity care, women and birthing people of color must also reckon with biases that have been embedded into medical algorithms and treatment guidelines. In the US, the Vaginal Birth After Cesarean (VBAC) calculator—an algorithm used for assessing the safety of a vaginal birth after a previous cesarean—contained specific adjustments for race and ethnicity until as recently as May 2021. These adjustments meant that Black and Hispanic women and birthing people were systematically assigned a lower chance of a successful VBAC, compared with their White counterparts. This is particularly concerning, given that successful VBAC has been shown to provide benefits to health, such as reducing the risk of bleeding and infection after birth and of complications during subsequent pregnancies.[40]

Race and ethnicity adjustments were initially included in the VBAC calculator in an attempt to account for research observations, which indicated that being White was associated with a higher chance of having a successful VBAC. A 2006 study investigating disparities in VBAC success between African American or Hispanic women and White women speculated that "ethnic variation in pelvic architecture may contribute." But as Darshali Vyas, a resident physician at Massachusetts General Hospital, and her colleagues pointed out in a scientific article in 2019, racial and ethnic disparities in VBAC success are driven by social inequality rather than racial or ethnic differences in biology. And automating these disparities by inserting them into a medical algorithm certainly isn't a solution. "If we systematize these existing disparities by building race/ethnicity subtraction factors into a predictive tool, we risk ensuring that these trends will simply continue," they wrote.[41]

In May 2021, William Grobman at Northwestern University in Illinois and colleagues—the researchers who developed the original VBAC calculator—published an updated version with the race and ethnicity adjustments removed. This was an important step, but the legacy of racial-ethnic adjustment in VBAC may continue to be felt for some time. According to the most recent National Vital Statistics Report on births in the US at the time of writing, which used data from 2019, the cesarean delivery rate remains higher among non-Hispanic Black women and Hispanic women, compared with non-Hispanic White women. Meanwhile, an earlier study of women who gave birth at the University of California, San Francisco, between 1990 and 2008 found that Black and Latina women were more likely than other women to have a cesarean delivery even after accounting for chronic diseases, complications of pregnancy, and sociodemographic factors.[42]

Meanwhile in the UK—shortly after the publication of the race-free VBAC calculator in the US—the National Institute for Health and Care Excellence (NICE) proposed draft guidance recommending that health care providers in England, Wales, and Northern Ireland consider inducing labor at thirty-nine weeks in pregnant women from "Black, Asian and minority ethnic backgrounds," due to their "higher risk of complications associated with continued pregnancy." The draft guidance prompted widespread criticism from doctors and patient advocates, who argued that offering early induction based on ethnicity wouldn't address the complex underlying factors contributing to poorer birth outcomes among ethnic minority women and could result in women feeling pressured to go down this particular treatment pathway without evidence that it would necessarily benefit them. In the final version of the guideline published by NICE in November 2021, the controversial ethnicity-based recommendation had been removed. The final guideline instead recommends considering induction from forty-one weeks for all uncomplicated pregnancies. Importantly, it also tells clinicians to be aware that "women from some minority ethnic backgrounds or who live in deprived areas" have an increased risk of childbirth complications, and that these women may benefit from closer monitoring and additional support.[43]

When I spoke with J'Mag Karbeah, the health services researcher at the University of Minnesota, she described the process of pregnancy and giving birth as "one of the most fantastic things the human body can do." As beautiful as that sounds, it also scares me: I am at a point in my life where pregnancy is shifting from something I have spent years actively trying to prevent, to something my partner and I think we might want. Both notions—not wanting to become pregnant and trying to become pregnant—carry several anxieties in my mind.

For as long as I can remember, I have had access to contraception and known that having an abortion would be an option available to me in case of an unwanted pregnancy or one that posed a danger to my health. I recognize that this is a privileged position to be in and it isn't one I take for granted. The WHO describes abortion as an "essential health care service" and yet, in June 2022, the US Supreme Court overturned *Roe* v. *Wade*—the landmark 1973 ruling that had provided a constitutional right to abortion. Since the court's decision, millions in the US have lost access to safe abortions in their states and, as ever, Black people and other people belonging to marginalized groups have been disproportionately affected.

According to 2019 data from the CDC, Black and Latina women are respectively four and two times more likely to have abortions than White women. In July 2022, the Kaiser Family Foundation reported that more than 4 in 10 women aged 18 to 49 living in states where abortion had already become or was likely to become illegal were women of color, including almost half of American Indian or Alaska Native women in that age range. It seems particularly cruel to me that people belonging to marginalized groups who are already disproportionately at risk of dying during pregnancy and childbirth are also often deprived of reproductive choice and access to this vital form of health care. And it isn't only the US where abortion access is being curtailed. Many anti-abortion organizations founded in the US now have branches abroad, including in Europe where I am based, and countries including El Salvador, Nicaragua, and Poland—where my partner is from—have also rolled back abortion access in the last few decades.[44]

When it comes to trying to become pregnant, I also have some concerns. Racial and ethnic inequalities are present at every stage of pregnancy, even before pregnancy begins. Statistics on miscarriage are sparse, but a 2021 study that analyzed data from seven countries in North America and Europe suggests that the risk is higher for Black women compared with White women. The analysis of data from 4.6 million pregnancies in the US, the UK, Canada, Sweden, Denmark, Finland, and Norway indicated that the risk of early pregnancy loss was 43 percent higher for Black women compared with White women. In the US and UK, research suggests Black women also face more barriers in accessing fertility treatment. Black women in the US undergo fewer in-vitro fertilization (IVF) cycles and have lower IVF birth rates compared with White women on average, despite being about twice as likely to experience infertility. In the UK, meanwhile, Black patients receiving IVF tend to be older than average on presentation and both Black and Asian patients tend to have lower IVF birth rates compared with White patients.[45]

In addition to inequalities in access to fertility treatment, other factors such as a lack of trust in health care services due to experiences of racism, and racist stereotypes about Black hyper-fertility, may also contribute to these disparities. "Such factors are largely unexplored within the fertility discourse, which has overwhelmingly focused on the experiences of White women," wrote Christine Ekechi, co-chair of the Race Equality Taskforce at the Royal College of Obstetricians and Gynaecologists, in a 2021 op-ed in the medical journal *The Lancet*. "Understanding how belief systems, fertility knowledge, and cultural and religious influences intersect with racism, access, and individual health factors is the only way to meaningfully bridge the gap in fertility outcomes," she said.[46]

Beyond becoming pregnant, the whole pregnancy journey itself seems quite daunting to me. Starting from the beginning of pregnancy, the list of things that could go wrong seems endless and overwhelming. When it comes to labor and birth I imagine that like many people, I would probably feel anxious about the pain of contractions, the potential medical interventions, and the overall uncertainty of the birthing process. In addition to these concerns, I know for a fact that—as with many other women and birthing people of color—I would also worry about racism. Jenny Douglas, the public health researcher and maternal health expert in the UK, told me that for that very reason she felt fortunate to have her mother present with her at the hospital during the birth of her first child. Her mother worked in the NHS as a midwife at the time and advocated for her during the birth.

"It's sad but that's the advice I give to everybody now, whatever they're going into hospital for, to actually take somebody with them, who can advocate on their behalf," said Douglas. This advice is pragmatic but, as Douglas pointed out, sad—and clearly not feasible for everyone. The Five X More website offers similar advice in terms of finding an advocate, as well as several other tips for Black women and birthing people in the UK that Douglas endorses, including to do research on pregnancy and labor, to keep a record of medical decisions and treatments, to speak up if something doesn't feel right, and to seek a second medical opinion where necessary. This goes some way toward assuaging my own personal worries about potentially going through all of this myself one day, but it is dispiriting that what is already likely to be an emotional and complex experience needs to be further complicated by the persistent racial and ethnic disparities that necessitate this kind of advice.

Naomi Williams was born in Tumut in New South Wales, Australia, a town which sits on the traditional lands of the Wiradjuri, Walgalu, and Ngunnawal Aboriginal peoples. According to her mother, Sharon, Naomi was an outgoing child and enjoyed the local Aboriginal community in Tumut and in Brungle, a nearby village. She was involved in community activities from an early age and loved movies, music, writing poetry, and painting. Naomi later moved to Canberra, Australia's capital, where she studied Aboriginal Art and Design at the Yurauna Centre, before eventually returning to Tumut because she missed her hometown and her friends. Back in Tumut, she started working as a support worker for disabled adults at a charitable organization called Valmar Support Services. She loved her work and was equally loved by her clients. She also had a loving relationship with her partner, Michael, and her stepdaughter, Kayla—and, in 2015, she discovered that she was pregnant. "We had the birth of our beautiful baby boy to look forward to," Michael would say later. "The dream Naomi and I wanted was starting to come together, where in life we wanted to be. It really was a dream come true."

Naomi had started feeling unwell early in 2015, before she became pregnant. She had been experiencing nausea and stomach pains, and visited Tumut Hospital at least eighteen times over the course of the next months. But Naomi wasn't referred to a specialist during any of these visits, neither before nor after she learned she was pregnant. Naomi's mother Sharon wrote to the hospital in July 2015 to complain about the way Naomi was being treated. She wrote that Naomi did use cannabis but that she was being "stereotyped as some sort of drug addict" and repeatedly referred to drug and alcohol services, which she said was "adding extra stress to her." Sharon said she was very concerned about

Naomi's condition and specifically raised the need for specialist referral. But Naomi's problems persisted, and she was not given access to a specialist.

Naomi had a difficult pregnancy and was eventually diagnosed with hyperemesis gravidarum, a severe form of morning sickness. By this point, she was so unhappy with the care she was receiving at Tumut Hospital that she was planning on returning to Canberra, where her mother was living, to have her baby there.

In the first minutes after midnight on New Year's Day in 2016, Naomi was driving herself to Tumut Hospital's emergency room. Her partner, Michael, later told an inquest that she had been feeling "quite unwell" the previous evening, that she "struggled to get up and get out of bed," and that she "began to burn up." She arrived at the hospital around fifteen minutes after midnight and was seen by a triage nurse and a midwife. She was assessed twice and given Panadol, before being sent home. She was discharged just over half an hour after she was triaged.

By the afternoon, Naomi was being brought back to the hospital in an ambulance. She had collapsed at home and become unresponsive. By the time she arrived at the hospital, her heart had stopped. Later, in a court statement in 2019, Naomi's mother, Sharon, said: "At the time of her death my daughter was a beautiful 27-year-old woman, passionate about social justice, excited about being pregnant with her first child and she was highly respected for the strong, hard-working Wiradjuri woman she was."

In July of 2019, an inquest into Naomi's death found that there was implicit racial bias in the way that she was treated at the hospital, which lowered her expectation that she would be treated well on the day she died. The cause of Naomi's death was identified as sepsis, a life-threatening reaction to an infection. Sepsis is an extremely serious condition but it is usually treatable with antibiotics if it is identified early. The coroner, Harriet Grahame, concluded that Naomi should have been examined further by staff when she presented to the emergency room in the early hours of New Year's Day in 2016.[47]

I came across Naomi's heartbreaking story and the findings of the inquest about a year later, at around the same time I learned of D'lissa's tragic experience in the UK in 2015. It struck me then that what had happened to D'lissa was connected to what had happened to Naomi on the other side of the world in Australia, and that both women's experiences were also connected to Serena Williams's experience in the US. In 2021, Naomi's cousin, the author and academic Anita Heiss, kindly connected me with George Newhouse, CEO of the National Justice Project, an independent human rights legal service based in Sydney that represents Naomi's family. During an intense conversation spattered with legal jargon, Newhouse talked me through the distressing details of Naomi's experience and highlighted why the coroner's findings in her particular case were so significant.

"It's the first time that a coroner in Australia has made findings that racial bias contributed to a death," Newhouse told me. "We were fortunate to have a coroner like Coroner Grahame, who was able to see the systemic racism in this case. More importantly, Naomi's mother, Sharon Williams, had the prescience to write to Tumut Hospital to identify and complain about the stereotypical and prejudiced care that her daughter was receiving. As a caring mother, she could see how her daughter was being harmed by the way her medical complaints were handled, and yet, the hospital ignored her pleas," he said. "So this was a rare case where prejudice was documented. It was only because of that objective evidence that systemic racism was ever going to be acknowledged."

Newhouse is right that the problem is systemic. At this point, the common thread of racism running through the experiences of Naomi and so many others around the world felt undeniable to me. Naomi wasn't even the first person in her family to experience racism at Tumut Hospital, I discovered. Both Naomi and her mother Sharon were born at the hospital, and Naomi's maternal grandmother, Nellie, was employed there as a domestic worker for about twenty-five years. "Her life there was tough. The hospital was historically a segregated hospital. It was only de-segregated not many years before Nellie started there and she was subjected to racial prejudice throughout her career although she developed many lifetime friends amongst the domestic staff," Sharon said in her statement to the inquest. "Because of my mother's long relationship with Tumut Hospital and our family friendships with staff there, I hoped and trusted that they would take on board my concerns and my daughter's pleas for care, but it seemed to me that the Hospital continued to treat Naomi in an uncaring way whatever we did or said to them."

These inequities also manifest at a national level in Australia. According to the Australian Institute of Health and Welfare, an independent government agency, the rate of pregnancy-related death among Aboriginal and Torres Strait Islander women is almost four times higher than that among non-Indigenous women. Indigenous women also have a greater risk of serious complications during pregnancy; a 2020 study found that the risk of severe maternal morbidity was almost twice as high among women of "Aboriginal," "African,"' or "Other" ethnicity compared with "Caucasian" ethnicity women. That study found that most of the disparity between "Aboriginal" and "Caucasian" ethnicity women couldn't be explained by taking into account sociodemographic characteristics and lifestyle factors such as advanced maternal age, lower socioeconomic status, and preexisting medical conditions.[48]

In a statement after the conclusion of the inquest into Naomi's death, Jill Ludford, chief executive of the Murrumbidgee Local Health District which covers Tumut, expressed her condolences to Naomi's family and friends and expressed a commitment to make improvements. "We are committed to making health care a more positive experience for our Aboriginal community," said Ludford. Naomi's family hasn't stopped seeking justice.[49]

Following the inquest into Naomi's death in 2019, the coroner made nine recommendations to the Murrumbidgee Local Health District, which it said it would review and consider, including strengthening a local Aboriginal liaison health worker program, adopting targets for employment and retention of Indigenous health care workers and auditing systemic bias. "Naomi's family and their relatives spent a lot of time crafting the recommendations about improving health care to First Nations peoples," explained Newhouse. "So the coroner's recommendations are, in very large part, the work of the family."

The coroner also made two recommendations centered around developing strategies for the provision of culturally safe or culturally appropriate health care. Culturally safe care is based on an understanding of and respect for culture and cultural difference, as well as an acknowledgment and appreciation of the historical context of colonization and its legacy of racism and inequality. The emphasis is on the person being cared for, including their experience of the care and their ability to access services and to raise concerns. The Congress of Aboriginal and Torres Strait Islander Nurses and Midwives (CATSINaM) has been advocating for cultural safety in health care across Australia for years. Two years after Naomi's death—and following consultation with CATSINaM and other organizations—the Nursing and Midwifery Board of Australia added the principle of cultural safety to its codes of conduct. This is significant, because evidence from Australia and elsewhere suggests that the provision of culturally safe and culturally centered care can improve health outcomes among marginalized racial and ethnic groups.[50]

To give an example: in Australia, maternity care based on principles of "Birthing on Country"—a more than sixty-thousand-year-old tradition of Aboriginal and Torres Strait Islander peoples giving birth on their ancestral lands—has been shown to significantly improve maternal and infant health indicators. A clinical trial, which followed more than 1,400 First Nations Australians who gave birth at Mater Mothers Public Hospital in Brisbane, Queensland, between 2013 and 2019, found that those who enrolled in a Birthing on Country service called Birthing in Our Community were more likely to attend five or more prenatal visits and less likely to give birth prematurely compared with those who received standard care. Birthing on Country models are generally First Nations–led maternity services that provide for the inclusion of traditional cultural practices and involve connections between land and country, but which can be incorporated in any setting.[51]

"Birthing on Country is not something new, it is a continuation of thousands of years of knowledge and practice," Janine Mohamed, former CEO of CATSINaM, told the *Australian Nursing & Midwifery Journal* in 2019. "Birthing on Country models of care provide integrated, holistic and culturally safe and respectful care for the best start in life for Aboriginal and Torres Strait Islander families and communities," said Mohamed, who is now CEO of the Lowitja Institute—Australia's national institute for Aboriginal and Torres Strait Islander health research, based in Melbourne.[52]

Increasing access to birth centers across rural and remote areas in Australia may help to provide more people with culturally safe maternity care, including Birthing on Country services. In Canada, the reestablishment of traditional midwifery and community birth centers in Inuit regions such as Nunavik, in northern Quebec, in the 1980s was linked to improved maternal health outcomes among Inuit women. The first such birth center was opened in the village of Puvirnituq in Nunavik in 1986, in response to community organization by Inuit women who were unhappy about contemporary policies of medical evacuation for births. Under those policies, women were frequently flown hundreds of miles south for hospital births in Montreal, and often ended up spending weeks or months away from home. As described in a submission of the Nunavik working group to Quebec's health ministry in 2003: "This intimate, integral part of our life was taken from us and replaced by a medical model that separated our families, stole the power of the birthing experience from our women, and weakened the health, strength, and spirit of our communities."[53]

The birth center in Puvirnituq was established following consultations with women and elders in the community, as well as traditional midwives. Its success led to the establishment of similar birth centers in other villages along Nunavik's Hudson Bay coast, such as Inukjuak and Salluit. The majority of the Hudson coast population now has access to maternity care within the region. In contrast, access to birth centers in remote parts of Australia, where Aboriginal and Torres Strait Islander peoples are more likely to live, is more limited. A 2017 study found that health facility catchment areas in which Aboriginal and Torres Strait Islander peoples made up more than 10 percent of the population were more likely to have reduced access to birthing services. This is despite research from Australia suggesting that planned birth center births among women with uncomplicated pregnancies are associated with positive maternal outcomes. And this isn't unique to Australia: an analysis of birth outcomes among women with low-risk pregnancies in several high-income countries found that planned birth center births were associated with higher odds of vaginal delivery and reduced odds of medical interventions such as cesarean, compared with planned hospital births.[54]

"The entire birth center model is that it's more patient-centered," Karbeah explained. She has done extensive research exploring the benefits of birth center birthing and culturally centered care for Black women and birthing people in the US. "Generally, we know that birth center birth folks usually feel more empowered, they feel more listened to," she said. Indeed, in their analysis of a culturally centered care model used by Roots Community Birth Center—an African American–owned, midwife-led freestanding birth center in Minneapolis, Minnesota—Karbeah and her colleagues at the University of Minnesota note high levels of patient satisfaction and point out that none of the 284 families served by Roots in the four years up to 2019 experienced premature births. "Roots is fantastic—and a place that I think should exist for everyone," said Karbeah. "But we know the reality is that it can't."[55]

There are several barriers limiting access to birth center births in the US, many of which disproportionately affect women and birthing people of color. Cost is a big one. Black, Hispanic, American Indian or Alaska Native, and Native Hawaiian or other Pacific Islander people of working age in the US are more likely than working age White people to be uninsured or to be recipients of Medicaid—public health insurance for people on low incomes. And despite a federal mandate requiring birth center services to be covered under Medicaid, a 2020 study found that in practice, reimbursement and other policy barriers often prevent birth centers from serving Medicaid recipients.[56]

These factors may explain why Black and Hispanic women in the US are less likely than White women to give birth in freestanding birth centers (although it's worth noting that birth center births represent a small minority of births in the first place, making up less than 1 percent of births in the US overall). This certainly doesn't seem to be due to a lack of demand. A California survey found that 14 percent of Black women, 12 percent of White women, 10 percent of Latina women, and 7 percent of Asian or Pacific Islander women said they would definitely want a birth center birth. Birth doulas, who provide emotional and practical support during pregnancy and childbirth, are also popular. The same survey found that Black and Latina women were more likely than average to say that they had received support from a doula during labor, although even greater proportions of women said they would definitely want doula support in the future, suggesting a significant unmet demand.

Increasing access to birth center births and doula services may go some way toward addressing inequities in maternal health, but this alone won't be enough to make disparities disappear. "I think we have to be very intentional about not letting hospitals off the hook," Karbeah told me. "It is very easy for a White hospital staff [member] to be like, 'Oh, Black woman, you're having a bad experience in our clinic? Why don't you go down the road to that Black birth center?'" she said. "But [. . .] not everyone's going to be low-risk enough to have a birth center birth. Not everyone necessarily wants to have a birth center birth. Some people feel more comfortable in the hospital setting. That's a personal choice."

Race and ethnicity are social constructs, not biological ones. But our race and ethnicity have a profound impact on our health, starting from the moment we first arrive in the world. Globally, people of color are not only more likely than White people to die while giving birth, they are also more likely to die while being born—or soon afterward. In 2018, the risk of a child dying before their first birthday was highest in the WHO African region, at 52 deaths per 1,000 live births. For comparison, the rate in the WHO European region was 7 per 1,000 live births. But racial and ethnic disparities also exist within nations, including in high-income countries such as the US and UK. Black infants in the US die at more than twice the rate of White infants, for instance. And across England and Wales, Pakistani, Black Caribbean, and Black African babies have the highest infant mortality rates, which are all around twice the rate among White British babies. Similar health inequalities in infancy as well as during early childhood have also been recorded in other countries, including in the world's two most populous nations. In western China—which is home to the majority of China's ethnic minority population—both maternal and child mortality rates are about twice as high among ethnic minority compared with Han Chinese populations. Meanwhile in India, rates of under-five mortality are higher than average among traditionally marginalized caste groups.[57]

Expecting parents have enough to worry about: a quick Google search reveals plenty of warnings about the risks of eating certain foods or taking specific medications during pregnancy, and the list of potential hazards only grows when it comes to infancy and early childhood. Yet the grave danger that racism poses to maternal, infant, and child health is too often overlooked. Racism drives social and economic inequality, limits access to culturally safe and high-quality health care, and disguises itself within medical textbooks as scientific fact, resulting in worse health outcomes among marginalized groups. And racism isn't only harmful during pregnancy and early life. Rather, it is an insidious threat to health, starting from birth and—as we'll see in the next chapter—persisting across the life course.

CHAPTER 2

Life and Death

Saidy Brown was born and grew up in Itsoseng, a small township in South Africa's North West province. The youngest of four siblings, Saidy had a happy childhood. "I was always the baby of the family," she told me. But in 2004, when Saidy was just nine years old, her family was hit by a tragedy. Her father suddenly and unexpectedly became seriously ill and passed away. Soon afterward, her mother became ill too, dying the following year and leaving Saidy and her siblings in the care of their aunt. The sudden loss of her father and then her mother was made even more painful by the mystery surrounding both of their deaths. Saidy had no idea what had made her parents so ill. "I just [knew] that they got sick, and they passed on," she said.

Saidy wouldn't learn what killed her parents until several years later, in 2009. That year, in June, she received life-changing news while hanging out with friends on a school trip to a Youth Day event. "We got there and there were people from this NGO, and they said, 'Guys, we are doing HIV testing for free.'" All around fourteen years old at the time, Saidy and her friends decided to get tested for fun. "I wasn't expecting my results to come back positive," Saidy told me. "I kept on saying, 'But I'm just fourteen, I didn't do anything,'" she said. "I also wanted to cry. But I thought that if I cry, the moment I walk out everyone will know my result. So, I didn't cry."

Ashamed and terrified about what her aunt and her peers would think, Saidy kept her diagnosis to herself. She had received no counseling at the time of her first positive test and believed that the diagnosis was a death sentence. To try to distract herself, she threw herself into her schoolwork and extracurricular activities. She joined a local drama club, which kept her busy with rehearsals after school. "For the next five months or so I was quite busy," said Saidy. "I didn't have time to be thinking about HIV or AIDS or death."

But it was unavoidable. As December approached, her drama club began working on projects related to HIV and AIDS in preparation for AIDS awareness month. "I kind of got triggered to tell someone," explained Saidy. She decided to tell her teacher, who immediately asked her if she had told her aunt. When Saidy replied, "No, you are the first person I'm telling," her teacher was very understanding and offered to come home with her so that they could talk to her aunt together. Saidy summarized how that conversation went: "We sat down, we told my aunt, and she's the one who said, 'I knew about your parents—I didn't know that you could have been born with HIV.'"

It would take another four years for Saidy to start antiretroviral therapy—a combination of medicines that treat HIV and prevent AIDS. Her aunt took her to a clinic as soon as she learned of Saidy's diagnosis, where her HIV status was confirmed, and she was given medicines to take home. But Saidy said it wasn't made clear to her that she would need to continue taking medication for the rest of her life. She didn't return to the clinic until several years later when she was eighteen. Since she hadn't been on treatment, her health had started to deteriorate. She had developed sores on her chest, which had begun spreading upward onto her neck. She had just started dating and was becoming self-conscious about her appearance.

"I was like, no, I'm trying to look pretty," said Saidy. "I don't want these sores on my face." So, she decided to return to the clinic, although she was apprehensive about taking antiretrovirals long-term because she had read about possible side effects on the internet. "I don't know what the page I landed on was, but this page said something like my hair was going to fall off, my teeth were going to fall off, my body was going to change. And I remember just going through the list and crying," recalled Saidy. "I've literally always been that person with the big hair. I was like, there's no way my hair's falling. And I love smiling, I love laughing," she said. "I texted my best friend and I said I'm just eighteen, I'm not doing this [. . .] These pills are just made to make me feel ashamed for living with HIV and to humiliate me. And I'm not about to subject myself to that. Rather I die."

Fortunately, on this particular visit to the clinic, Saidy received counseling. The counselor tried to reassure her that the side effects were not as drastic as what she had read online. Somewhat reluctantly, Saidy agreed to try taking the medication for a month. "Fortunately for me, I really started seeing changes," she said. The sores began to disappear, and she started to feel better. "Years later, I'm here and I'm still on treatment," she told me proudly when we spoke during AIDS Awareness Month in 2021.

Saidy's experience getting diagnosed with HIV as a young Black girl in South Africa and her journey accessing treatment exemplify long-standing and persistent racial disparities associated with what for years has been a completely preventable and treatable infection. These inequities exist not only in the context of infectious diseases like HIV, but also across several of the other most prevalent health conditions affecting humanity today. From infectious diseases to non-communicable health conditions such as cardiovascular disease, cancer, and mental illness, there are significant racial and ethnic inequities in terms of who is getting ill and who is dying. The racial and ethnic gaps in maternal and infant health are just the tip of the iceberg. As I discovered, gaps in physical and mental health exist throughout life—and illness is far from the great equalizer it is often touted as being.

Saidy is now an award-winning HIV activist who shares her experiences of living positively with HIV through videos and posts on social media. And while she is grateful that she has access to lifesaving treatment, she is painfully aware that this isn't the case for everyone. "Fortunately for me, I started treatment at a time when treatment had already been made free in South Africa. I'm also aware that that's not the experience of every person in the world," said Saidy. "The access aspect of things was not that difficult for me. But also, we don't know what happened with my parents," she added. "That's probably why I really got born with HIV, because there was no treatment [accessible to them] at the time."

The US Food and Drug Administration (FDA) approved the antiretroviral drug AZT, the first treatment for HIV, in 1987. In 1994, the year before Saidy was born, the US public health service recommended that pregnant women with HIV be given AZT to reduce their risk of transmitting the virus to their babies. By 1996, the year after Saidy was born and almost a decade before her parents died, antiretroviral therapy became the standard treatment for HIV—for those who had access to it. The combination of antiretroviral drugs quickly became accessible to people in the world's wealthier—and generally Whiter—countries, resulting in drastic reductions in deaths from AIDS in those nations. But access to the lifesaving treatment in the world's less wealthy nations lagged, particularly in countries in sub-Saharan Africa where access to treatment didn't become widespread until about a decade later because of the high cost. This is a persistent pattern: in 2023—more than two years after the world's first Covid-19 vaccination campaigns began—there remained significant global inequities in Covid-19 vaccination coverage, with Africa being the least vaccinated and least boosted continent.[58]

In South Africa, the lag in the rollout of antiretrovirals around the turn of the twenty-first century was exacerbated by Thabo Mbeki's presidency, as he embraced HIV and AIDS denialism and implemented policies that restricted access to the drugs. It has since been speculated that this appetite for denialism may have been fueled in part by a general suspicion of Western medicine as a result of the actions of the apartheid regime. Following the collapse of apartheid, it was revealed that—under the leadership of a medical doctor—apartheid government laboratories had been developing chemical and biological weapons to use against enemies of the regime, as well as researching methods of covert and selective fertility control. This example barely scratches the surface when it comes to Western medicine's racist history and legacy worldwide, which we will examine more later in the book.[59]

Today, the majority of people living with HIV live in Africa, and whereas the prevalence of HIV is about 0.2 percent worldwide, in sub-Saharan Africa it is about 6 percent. As of 2020, South Africa has the world's largest HIV epidemic with more than seven million people and almost 20 percent of 15- to 49-year-olds living with the virus. HIV prevalence is higher among Black South Africans, compared with their counterparts from other racial groups, and particularly among Black women, like Saidy. "It is something that really affects adolescent girls, young women, Black people, people who are less privileged," she told me.[60]

Even within wealthier nations, such as the US, there were stark inequalities in access to HIV treatment from the outset. In 1996, AIDS ceased to be the leading cause of death among all US adults between the ages of 25 and 44 but it remained the leading cause of death among African American adults in the same age range. Many years later, although the overall prevalence of HIV has fallen significantly since the 1990s, racial inequality persists. About 13 percent of the US population is Black, but as of 2019, Black people make up 40 percent of people living with HIV. And despite the higher prevalence of the disease among Black people, they are still less likely than White people to be receiving treatment and to have undetectable levels of the virus in their blood. Black people in the US are also six times less likely than White people to be prescribed pre-exposure prophylaxis or PrEP, which can prevent HIV transmission, and health care providers are less likely to discuss PrEP with Black clients.[61]

These inequalities are compounded for people with multiple intersecting, marginalized identities, such as Black transgender women. An estimated 14 percent of all transgender women in the US are living with HIV, but among Black transgender women the equivalent figure is estimated at 44 percent. Similar inequalities exist in other multi-ethnic societies globally. For instance, a 2021 study in the UK found that people from Black, Asian, and minority ethnic groups with HIV tend to present to health care services with poorer immune health, spend less time engaged in care, and are more likely to experience a rebound of the virus.[62]

This affects us all—racial inequities are holding back efforts to halt the spread of HIV and end AIDS globally. In 2022, the UN warned that progress in prevention and treatment was slowing worldwide, noting that racial diagnostic disparities were exacerbating HIV risks and pointing out that declines in new HIV diagnoses had been greater among White populations than among Black and Indigenous populations in countries such as the UK, the US, Canada, and Australia.

In a sad irony, racial and ethnic inequalities in HIV prevalence contribute to racism and racialized HIV stigma, which discourages people from getting tested and taking treatment where it is available to them, perpetuating existing inequalities further. I will never forget the moment when I was doing some outreach work at a school in the Oxfordshire area in the UK, talking to seven- and eight-year-olds about my research on viruses and the immune system, when one of the kids suddenly shouted, "HIV comes from Africa because African people are dirty." I felt taken aback and disturbed. In her book *Divided*, UK-based sexual and reproductive health doctor Annabel Sowemimo recalls feeling similarly unsettled after she was approached by a young Black gay man at an event she attended with her nonprofit organization Decolonising Contraception, who told her he had stopped using dating apps after several users had refused to engage with him due to fears that "Black men have HIV," as one user told him.

Racist, homophobic, and stigmatizing perceptions like these are not only morally offensive, they actually harm people's health. As Sowemimo goes on to illustrate in her book, racialized stigmatization and stereotyping contributes to wider inequalities in sexual and reproductive health, by discouraging those affected from engaging with health services. Health harms can be particularly acute for those who are living with HIV. According to a 2020 UNAIDS report, between about 2 and 21 percent of people living with HIV across 13 countries with available data reported being denied health care in the previous 12 months because of their HIV status. Separate research has linked HIV stigma to low rates of HIV testing and lack of knowledge about HIV treatment.[63]

Saidy said she thinks stigma and racism influenced her decision to keep her diagnosis to herself for several months and to delay seeking treatment for years. "I literally felt like people were going to judge me," she said. "I do think that racism and HIV stigma overlap."

As an HIV activist, Saidy now works hard to fight against discrimination, stigma, and misconceptions about people living with HIV. In an anecdote she shared with me about some time she spent volunteering with a student activist organization, she highlighted how racialized stigma harms people of all races. "They [the organization] did a lot of HIV advocacy on campus. And we would have these wellness weeks where we would test students. And we would have these White boys who would feel like, 'No, I can't even test for HIV. Why would you ask me? I don't even look HIV positive,'" explained Saidy. Anything that discourages people from getting tested is harmful. "HIV affects everyone," she said, regardless of race.

Although the stigma surrounding HIV is particularly pronounced, it is far from the only infectious disease to be stigmatized and connected to discrimination against particular racial or ethnic groups. The UK, the US, and Australia were among countries that saw surges in anti-Asian hate speech or other racist incidents within the first two years of the Covid-19 pandemic. And in November 2021, as reports emerged about the detection of the omicron coronavirus variant by researchers in Botswana and South Africa, the Spanish newspaper *La Tribuna de Albacete* published a comic depicting viruses as cartoon characters with dark skin and Afro hair, packed into a boat labeled with a South African flag and heading toward land marked with a European Union flag. WHO director-general Tedros Adhanom Ghebreyesus criticized the depiction in a tweet, writing: "It pains me that shows of racism like this still plague the challenges facing the world today. Caricaturing people crammed in a boat bringing a virus to Europe is disgusting. We can only advance, as one [global] community, by promoting solidarity, not stigma." I was similarly disappointed, as a Black person living in Germany, to see German newspaper *Die Rheinpfalz am Sonntag* publish a front-page article on the same day with the headline "The virus from Africa is with us," above a photograph of two Black people. Both newspapers subsequently apologized. Yet in May 2022, the Foreign Press Association Africa felt compelled to issue a statement criticizing European and North American media outlets for repeatedly using images of Black people to illustrate articles about mpox outbreaks in Europe and North America. (The WHO renamed the disease from monkeypox to mpox in November 2022, saying that the original name played into "racist and stigmatizing language.")[64]

Unfortunately, there is often stigma associated with infectious diseases, particularly in the context of emerging epidemics or pandemics where there are many unknowns and fears. The 2013–2016 Ebola virus epidemic in West Africa was also associated with significant stigma, for instance, including racialized stigmatization of African immigrant populations in other parts of the world, such as the US. And as we have already seen with HIV, it isn't only stigmatization that affects marginalized racial and ethnic groups disproportionately during epidemics or pandemics—these groups often also experience disproportionate rates of disease and death.[65]

This latter trend was impossible to miss when the world was gripped by the Covid-19 pandemic in 2020. It was around this time that I first came across Nishi Chaturvedi, a clinical epidemiologist at University College London in the UK, who has done extensive research on ethnicity and health. In May 2020, Chaturvedi and I had both been invited as guests on the VICE UK podcast, *VENT Weekly*, to speak about why people from Black, Asian, and minority ethnic backgrounds in the UK were disproportionately getting ill and dying with Covid-19. In introducing the discussion, our adept podcast host, Amelia Mya, touched on a question many people were asking at the time: *Is the coronavirus racist?* "I do feel like a lot of people have said that this virus doesn't discriminate in terms of race, but I have a feeling that it does," said Mya.

She had a valid point. A few weeks before the podcast recording, one of my editors at *New Scientist* had asked me to investigate and report on emerging data on the race and ethnicity of Covid-19 cases in the US and UK. Figures published by the UK's Intensive Care National Audit and Research Centre as early as April 2020 were already showing that people from Black, Asian, and minority ethnic backgrounds were overrepresented among critically ill Covid-19 patients relative to their share of the population. Data published by the CDC in the US around the same time revealed a similar pattern, with Black people particularly overrepresented. On April 18, 2020, for instance, when I was preparing my report, 30 percent of confirmed Covid-19 cases where race was known had been found to be among Black or African American people, who comprise just 13 percent of the US population. Similar inequalities were also detected in other countries that collected the relevant data, such as Australia and Brazil.[66]

These patterns of inequality were unfortunately nothing new. "This isn't a new phenomenon that's specific to Covid-19—it's something we've seen before," commented Chaturvedi at the time. Indeed, ten years before Covid-19, the world was hit by the H1N1 influenza virus pandemic, also known as swine flu. That virus was a lot less deadly than the coronavirus responsible for Covid-19, but there were similar inequalities in terms of who died. A study in England found that people of non-White ethnicity had nearly twice the risk of dying compared with White people, for instance.[67]

Similar racial and ethnic disparities exist for tuberculosis, or TB, one of the oldest diseases affecting humanity. "I think diseases like TB, like

Covid-19, invariably they hold a mirror to us and explain to us the overt racism and the inequities that are just rampant. They've always been there," said Madhukar Pai, an epidemiology and global health researcher at McGill University in Canada, who studies TB.[68]

In a similar vein, back on the podcast, Chaturvedi told listeners: "When Covid goes away this story will remain, that people who are disadvantaged in some way are at greater risk of some conditions. So, I'm hoping that one of the positives that will come out of this is that recognition—and that will play through into government policies and communities and how we live our lives." Indeed, as we will see when we return to the issue of Covid-19 inequalities in the next chapter, it isn't viruses or diseases that are racist—it is our societies. Racism in societies damages health overall, leaving people more vulnerable to infectious assaults. And, as we are about to learn, one of the major ways this damage manifests itself is on the heart and circulatory system.

In January 2018, Carol Ighofose was wrapping up a shift at her work as a GP, or family doctor, in the Leicestershire area of England, when she began to feel a slight discomfort in the upper part of her stomach—like heartburn. It's probably a bit of indigestion, she thought to herself, as she prepared to drive home. She had been fasting for religious reasons, so she made a mental note to herself to eat something when she got home and before returning to work for her next shift later that afternoon. Carol got into her car and called her husband, Simon, to let him know that she was on her way home. She then called one of her friends, Tracey, using the hands-free Bluetooth system in the car, so that she could chat to her during the drive home. Tracey, who was working as a GP trainee at the time, had finished a night shift earlier that day but had stayed up because she had wanted to speak to Carol about something. As they were chatting, Carol realized that her indigestion symptoms still hadn't settled.

"Trace, you know, I don't usually get chest pain but this discomfort in my chest that I have been experiencing is not shifting," Carol told her friend. Both doctors, they went straight into diagnosis mode. "[Tracey] started asking me all the questions about, you know, what the pain is like, how would I grade it, how would I describe it, all those sorts of things," Carol recalled a few years later. They concluded that something didn't quite add up and that if Carol's symptoms persisted, she should go and see a doctor. "At this time, even though we were thinking heart, and she did ask all of the heart things, it just didn't add up," Carol explained. "Thinking about the risk factors, I wasn't overweight, I wasn't diabetic, I didn't have high blood pressure, my cholesterol was normal, I did my exercises every morning before I'd go to work. It just didn't fit."

But as Carol continued driving, her symptoms worsened. She was still on the line to Tracey but was struggling to concentrate on what her friend was saying, and the discomfort she had been experiencing had developed into more of a chest pain. "This is not right, I need to pull over," Carol told her friend.

"Pull over, I'll call 999," Tracey replied.

As Carol pulled over on the side of the road, and her symptoms continued to worsen, it dawned on her what was happening. "At this time, in my head, I realized I'm having a heart attack," she said.

Carol waited in her car for what felt like an eternity for the paramedics to arrive. By this point, she was experiencing crushing central chest pain and was feeling extremely disoriented and confused. She decided to call 999 herself. "I am a doctor and I believe I'm having a heart attack," she told the operator. She was relieved when the paramedics arrived about ten minutes later, but she was surprised that they appeared to be quite laid-back. "In my head, I'm having a heart attack, so I expected when they came that they would be like on mission, urgent, but they were quite lackadaisical," she recalled later.

"They parked the ambulance a little way down from where my car was and then came to me and got me to secure my car, i.e., wind up the windows et cetera, and then walk to the ambulance from where my car was. I don't know how I did all of that, because I was in a disoriented state, and that in itself told me these people aren't taking me serious. If you've got a heart attack, you don't want the patient to walk, you want the patient to rest as much as possible," Carol explained, slipping into doctor mode. Even as she was being walked to the ambulance from her car, experiencing what must have been excruciating pain, Carol had her doctor's hat on. She remembers emphasizing to the paramedic, "I am a doctor, I do believe I'm having a heart attack." The paramedic just looked at her and said, "Oh, you doctors always think of the worst."

In the ambulance, the paramedics did start the correct treatment for a heart attack, according to Carol. "They did give me an aspirin, they started putting a vent flow in my vein for venous access so I could get fluids," she remembered. She was also given morphine and some medication to treat nausea and the paramedics did an electrocardiogram, or ECG, to measure the electrical activity of her heart. "What they said they found was what we call a high takeoff in medicine," said Carol. "That could be the first sign of a heart attack," she explained, so she expected that she would be taken straight to a heart specialist hospital for treatment. "If I were a GP, having me as a patient in front of me, I would want the patient to go to the heart specialist hospital which we have in Leicester where I'm from. So, I asked them to take me there," she said. But her ECG result was dismissed as something else, and Carol was instead taken to the nearest hospital, which was about five minutes away.

"Unfortunately, I went to that hospital, waited there for fourteen hours, even though I was constantly in pain—they had to keep giving me morphine," she remembers. Then, at around four o'clock the following morning, Carol was finally moved to Glenfield Hospital, the specialist hospital where she had asked to be sent initially. When she got there, the doctors confirmed her worst fears. "It was confirmed that I did have a heart attack," she said solemnly. "I had a total blockage of my left anterior descending artery," she added, slipping back into doctor-speak to refer to a critical blood vessel that usually supplies oxygenated blood to the left side of the heart. Blockage of the left anterior descending artery is extremely dangerous: it is sometimes referred to informally as a "widow-maker" because it carries such a high risk of death (and presumably also because, well, sexism). Despite Carol's serious condition, it took a further six hours—a total of twenty hours from the time she first recognized she was having a heart attack—before she had a stent put into her artery to relieve the blockage.

The next day, an ultrasound revealed that Carol had experienced significant damage to the left side of her heart. The amount of blood that her heart was able to pump out with each beat was extremely low. "Usually, we need about 65 percent of blood to perfuse all our body. Mine was pumping out 18.18 percent," Carol explained. "As a doctor, when I heard that I started crying, because for me that meant heart failure and a life of—a different life, basically."

When I heard Carol share her experience at a British Heart Foundation event in April 2021, part of me was shocked. This woman was a doctor, who had somehow had the presence of mind to diagnose herself and work out what kind of treatment she should be getting *while she was having a heart attack*, and yet she had still somehow been doubted by the medical professionals who were caring for her. Another part of me wasn't that surprised: it was unfortunately yet another example of a Black woman being disbelieved and dismissed about what was happening inside her own body.[69]

Carol believes that the delay in diagnosing and treating her heart attack meant that she suffered damage to her heart that might not have occurred had she received more timely treatment. She finished her talk with a message: "I'm grateful to be sharing my story and I hope that women in particular and people of ethnic minority and whoever—you know your body, if you think something is different, do, I always say, be nice, be polite, but please insist, because we know when something is not right with us."

Unfortunately, knowing our own bodies and being nice and polite sometimes isn't enough. Nishi Chaturvedi, the clinical epidemiologist whom I had met a year earlier on the podcast about Covid-19, was also a speaker at the same British Heart Foundation event, and during her talk, she illustrated why.

"Getting care for cardiovascular disease is a complex process that involves both the patient and the health care system," said Chaturvedi. "So, for example, if we're thinking about something like a heart attack, it requires the individual to recognize they're having an event, recognize that it's serious, choose to seek help and to seek help immediately, and to accept the interventions offered. The doctor has to make the correct diagnosis, do the right tests, perform the right investigations and provide the right treatments. And we do see some evidence that people from ethnic minorities experience delays in getting appropriate health care," she said. Chaturvedi and her colleagues have put a huge amount of effort into investigating why people from ethnic minority backgrounds in the UK, and particularly people of South Asian ethnicity and people of Black African or Black Caribbean ethnicity, such as Carol, not only experience a greater risk of cardiovascular disease compared with White people but are also more likely to experience delays in getting the urgent care that they need.[70]

"One of the studies we did was to send out a questionnaire to people of South Asian and European descent—middle-aged folk—and to give them a vignette, a story, of an individual walking upstairs carrying a heavy load and experiencing central chest pain, putting down the load and feeling a bit better, but concerned because they'd had that event before. And we asked our respondents, what do they think the pain is? Do they think it's serious? Would they seek help? Would they seek help immediately? And would they accept an intervention if it was offered? And, overwhelmingly, the South Asians were equally likely to recognize that this pain was coming from the heart as the Europeans and, in fact, even more likely to wish to seek help urgently and to accept interventions, be they medical or surgical," said Chaturvedi.

Chaturvedi then pulled up a PowerPoint slide with images showing the so-called standard symptoms of a heart attack. "When I was at medical school we were taught how to spot when someone had a heart attack, and the symptoms you see on the left hand of the slide here are the kind of things we were taught to look out for," she explained. "Chest pain, shortness of breath, the pain radiating up to the jaw, the neck, or down the shoulder—so that's what we expect to see when someone's having a heart attack," she said. Chaturvedi continued: "It's now clear that there are non-standard symptoms, things like dizziness, heartburn, cold sweat, unusual tiredness, and there's evidence to suggest people with diabetes, women, and ethnic minorities are more likely to present with these nonstandard symptoms."

Rewatching Chaturvedi's talk later, I thought back to Carol's experience. The first symptom Carol noticed that day when she was preparing to leave work was heartburn, but even though she is a doctor—and was on the phone to her friend at the time who is also a medical professional—it wasn't immediately obvious to either of them that this could be a heart event. Perhaps if there was more awareness about these so-called "nonstandard" symptoms of a heart attack, it would have been easier to work out what was happening to Carol. Even as Carol's symptoms persisted, she and Tracey initially dismissed the possibility that it might be a heart attack because none of the common risk factors—being overweight, being diabetic, having high blood pressure, or having high cholesterol—seemed to apply to Carol, who was just forty-eight years old at the time and led an active lifestyle. At a health checkup eight years before her heart attack, she had even been told that she could expect to live well into her eighties without having a heart attack or stroke.

In fact, both Carol's experience and Chaturvedi's research suggest that the standard symptoms and risk factors for cardiovascular disease might not be so standard after all. "These guidelines [for the diagnosis and treatment of cardiovascular disease] have largely been developed and deployed in people of White European origin and may not work so well with people of South Asian and African-Caribbean origin where [. . .] the constellation and interplay of risk factors is very different. Those guidelines do try to take account of different ethnicities, but our work suggests that they don't do it particularly well and that perhaps some ethnic minority groups are being undertreated," Chaturvedi went on to explain. "In terms of health care, the standard symptoms and guidelines might miss diversity in expression of symptoms and might miss diversity in risks of cardiovascular disease and interplay of different risk factors." This is not to say that people of different ethnicities have inherently different biology, emphasized Chaturvedi, but environmental risk factors can impact our biology.

We can see how this happens by considering the relationship between ethnicity, environment, and biology in cardiovascular disease. People of Black and South Asian ethnicity in the UK are disproportionately affected by type 2 diabetes, which is not only a major risk factor for cardiovascular disease but is also associated with an increased likelihood of experiencing some of those "nonstandard" heart attack symptoms we considered earlier. And although more than 120 gene variants have been identified as being associated with an increased risk of type 2 diabetes, research by Chaturvedi and her colleagues suggests that it is environmental factors, such as socioeconomic status, which make the strongest contribution in accounting for ethnic differences in type 2 diabetes risk.[71]

In a 2022 study, Chaturvedi and her team analyzed data from the UK biobank, a large-scale biomedical database containing genetic and health information from approximately 500,000 people aged between 40 and 69. Their analysis confirmed findings from previous studies that the incidence of type 2 diabetes among people of Black and South Asian ethnicity is higher than among White people in the UK. If this difference was mostly due to genetic differences, we would expect that it would remain largely unchanged between generations. But instead, the team found that among Black and South Asian ethnic groups, second-generation migrants to the UK had a 20 percent lower risk of type 2 diabetes compared with first-generation migrants. This was associated with reductions in socioeconomic deprivation and measures of body fat between generations.[72]

As Chaturvedi put it: "The genetic makeup hasn't changed that much but the environment has." These results are hopeful, as they suggest that the persistent gaps in type 2 diabetes incidence between Black and South Asian ethnic groups and White ethnic groups can similarly be reduced by focusing on environmental risk factors such as ethnic disparities in access to healthy food and to green space for exercise, which are both enormously protective against obesity, type 2 diabetes, and cardiovascular disease.

In the next chapter, we will examine in more detail how systemic racism contributes to racial and ethnic inequalities in diet and exercise in the UK and elsewhere, but right now I want to return to Carol. When she attended her health checkup eight years prior to her heart attack, she didn't have any of the risk factors we have just been exploring. She wasn't obese and she didn't have diabetes. But she had been experiencing something else, which may have adversely impacted her health. "I think looking back, for me personally, of course I'm a busy GP and my life is generally, from that perspective, stressful and there are other stresses, so I think that stress does play an important role," reflected Carol, during a Q&A session after her talk. Putting her medical hat back on, she added: "Stress is not listed as an independent risk factor for heart attacks, but it's listed more as an association. I personally feel that some more work needs to be done on that."

Carol's speculation about the potential impact of stress on her heart reminded me of research I had come across previously, exploring the connections between racism, chronic stress, and cardiovascular disease. To understand these connections, it helps to remind ourselves of our basic plumbing.

The heart is one of the most amazing organs in the human body. It is also kind of a glorified pump, channeling oxygen-rich and oxygen-poor blood to and from the body's organs with every beat, through a spectacular network of fleshy pipes. Those pipes are the unsung heroes of this circulatory dance—a fact that becomes obvious when one becomes blocked, and even more startlingly so when the blockage happens in a vessel supplying a crucial organ such as the brain or the heart itself. Indeed, strokes and heart attacks are mainly caused by blockage of blood flow to the brain or heart, most often due to a buildup of fatty deposits in the inner walls of the blood vessels that supply these organs. These buildups tend to happen in regions of the blood vessel wall that have been damaged for some reason. Once damage has occurred, substances traveling in the blood, such as cholesterol and fat, may accumulate inside the damaged area and trigger a runaway cycle of inflammation and further damage. This process can culminate in the formation of a plaque—a chaotic lump of gunk that bulges out into the blood vessel, restricting the flow of blood and potentially resulting in the formation of dangerous blood clots that can block blood flow through the vessel entirely. Having high blood pressure significantly raises the chances of all these things happening, because the large force exerted by the circulating blood increases the likelihood of damage to the blood vessel walls. It also puts strain on the heart. As a result, hypertension—consistently elevated blood pressure—is one of the biggest risk factors for cardiovascular disease.

Chronic stress has been identified as one of a number of factors, also including genes, unhealthy diet, and physical inactivity, which can contribute to the development of hypertension. We have all experienced those telltale physical sensations associated with increased heart rate and blood pressure during a stressful situation. These effects are driven by a rush of hormones, such as cortisol and adrenaline, which usually subside once the stressful situation ends. The problem comes when the stresses—and the associated hormones and elevated blood pressure—don't end.[73]

Over the past few decades, a number of studies have identified racism as a chronic stressor that may contribute to hypertension and negatively impact cardiovascular health. We will explore this research in more depth later in the book, but broadly, perceived discrimination and racism-related vigilance—an anticipatory stress associated with living in a racist environment—have both been associated with hypertension prevalence among Black people in the US, among whom the incidences of hypertension and cardiovascular disease are higher than those among the wider US population. In the UK, the incidence of hypertension is also higher among people of Black ethnicity, as well as among people of South Asian ethnicity, relative to that among White ethnic groups. And although Carol's blood pressure had appeared normal at the time of her health checkup, she had experienced high blood pressure in the past.[74]

"I had in both my pregnancies, pregnancy-induced hypertension," a form of high blood pressure in pregnancy, she recalled. "Looking back, I have discussed and I've wondered whether that played any role at all," she added. Hypertensive disorders in pregnancy are a leading cause of pregnancy-related death, which as we saw in the last chapter is more common among Black women than White women in both the US and UK, and there is evidence that women who experience hypertension during pregnancy have an increased risk of developing cardiovascular disease later in life.[75]

From listening to Carol speak, it is clear that she is a person who takes a lot of responsibility for her health—and her medical background means that she approaches everything with clinical precision. She explained that after her heart attack, she looked back through her medical records and noticed that although she hadn't been diagnosed with diabetes before, her blood glucose had been slightly above normal in the past. Since the heart attack, she has been struggling to get her blood glucose level down. "I've lost even more weight, my lifestyle is—I don't know what more I can do, but those things," she said.

I sympathized with Carol's need to understand the factors that led to the serious and life-altering damage sustained to her heart. Was it her "nonstandard" symptoms? Was it her being disbelieved when she sought treatment? Was it stress? Like Carol, we can't know for sure. But her experience points toward a nexus of factors, many of which have racism at their center. This has broad implications: cardiovascular disease is the world's biggest killer, even though most cardiovascular diseases are preventable by addressing environmental risk factors. As we will see in the next few chapters, improving access to healthy living conditions, equitable medical care, and an environment free from the stresses of daily discrimination would all go a long way toward reducing racial and ethnic inequalities in cardiovascular disease. This applies to other diseases too, including the world's second biggest killer—a disease that can take hold almost anywhere in the body.

I have never been especially interested in superhero movies or comics, even as a child. But in February 2018 when the first *Black Panther* movie started showing in UK cinemas, I absolutely could not wait to watch it. I finally went to see it one weekend with a friend in Oxford, and the movie more than lived up to my expectations. I remember feeling a bit emotional and overwhelmed. It was a movie I wished I could have watched as a child, growing up and going to school as one of few Black kids in a predominantly White environment. I used to feel ashamed of my skin, my hair, my general Blackness and "otherness," but I think watching that movie—a blockbuster with a Black director featuring a star-studded, mainly Black cast wearing dashikis and kente cloth and natural hairstyles—would have helped my child self to feel seen. This memory of watching *Black Panther* came flooding back to me two years later, in the summer of 2020, when I heard the news that the actor Chadwick Boseman—*the* Black Panther—had died from cancer at the age of just forty-three.

"It was the honor of his career to bring King T'Challa to life in Black Panther," his family said in an emotional statement shared to his Instagram account shortly after his death. Boseman had been diagnosed with stage III colon cancer in 2016, which later progressed to stage IV. His family revealed this in their Instagram post, to the shock of Boseman's fans, and explained that he had privately been undergoing countless surgeries as well as chemotherapy treatment, all the while continuing filming for multiple movies.

Boseman's death shattered so many people who, like me, saw him as a literal superhero. In a year during which Covid-19 had already exposed vast racial and ethnic health inequities in the US, the UK, and elsewhere in the world, his death also put a sharp spotlight on how these inequities manifest in cancer—the second biggest killer globally after cardiovascular disease. Just as cardiovascular disease encompasses a range of different conditions, cancer also refers to a diverse group of diseases. These diseases can appear almost anywhere in the body, because of the key feature that connects them: uncontrolled cell growth.

Usually, our cells grow and multiply through a meticulously controlled process called cell division (the first time I had the privilege of looking down a microscope and witnessing a cell in the midst of this highly choreographed routine, I was in awe). Cancers can develop in cases where this orderly process has broken down due to genetic mutations—changes to DNA—which are rare, but which tend to accumulate with age. Environmental exposures play the lion's share of the role in influencing the likelihood of these changes occurring, although they can also sometimes be inherited. Cancer cells look a lot like normal cells, but they can behave very differently: growing when they haven't been instructed to, ignoring signals to stop growing, spreading throughout the body, persuading blood vessels to grow toward them, and hiding from the body's immune system. The sooner these dangerous cells are detected, the better. Boseman's cancer was diagnosed at stage III, meaning it had most likely already started to spread from his colon to invade nearby parts of the body. When he died, within less than five years of his diagnosis, the disease had already progressed to stage IV, indicating that the cancer had disseminated further.

Boseman's experience unfortunately reflects a broader pattern in the US, which his wife, Taylor Simone Ledward, highlighted during a heartfelt speech she gave as she accepted her late husband's NAACP Image Award in 2021. "He was an uncommon artist and an even more uncommon person, but the manner in which we lost him is not uncommon at all—not in our community," said Ledward, fighting back tears. She was referring to the fact that Black people in the US are 20 percent more likely to develop colorectal cancer and 40 percent more likely to die from it, compared with most other racial and ethnic groups. In fact, Black people experience higher death rates and lower five-year survival rates for most cancers compared with other racial and ethnic groups in the US. This is particularly striking given that the incidence of cancer overall is similar if not slightly higher among White people compared with Black people, with White people experiencing the highest rate of cancer of any racial or ethnic group in the country.[76]

For instance, despite a similar incidence of breast cancer—the most common type of cancer—among Black and White women, the death rate from the disease is 41 percent higher among Black women. This seeming paradox is at least partly explained by disparities in cancer diagnosis and access to timely treatment. Black people are more likely than White people to be diagnosed at an advanced stage of disease—as Boseman was—by which point treatment is usually more expensive and less likely to be successful. Ledward addressed this during her speech, emphasizing the value of routine screening in catching colorectal cancer before it's too late. "If you are forty-five years of age or older, please get screened. Don't put it off any longer," she said. "This disease is beatable if you catch it at its early stages, so you don't have any time to waste," she added.

While efforts to encourage people to get screened for cancer regularly are laudable, this alone won't be enough to eliminate disparities in disease outcomes. It will also be crucial to make certain that access to screening is equitable in the first place. Black people and other people of color in the US are more likely than White people to be uninsured and to face other barriers that may limit access to health care, including routine cancer screening. And even beyond diagnosis, there is evidence that Black people and people from other racial and ethnic minority backgrounds are more likely to experience delays in their cancer treatment relative to their White counterparts.[77]

At this point it probably won't surprise you to learn that racial and ethnic disparities in cancer have also been documented elsewhere in the world. In 2021, New Zealand politician and member of parliament Kiri Allan shared her cervical cancer diagnosis on social media, in an effort to encourage more people to get screened. "Please, please, please—encourage your sisters, your mothers, your daughters, your friends—please #SmearYourMea—it may save your life—and we need you right here," she wrote, using a slogan created by the late cervical cancer campaigner Talei Morrison to encourage Māori women in particular to get screened for the condition.

The vast majority of cervical cancer cases are caused by infection with human papillomaviruses, or HPV, and the combination of HPV vaccination with regular cervical screening has been shown to significantly reduce the risk of disease and death. In 2020, about 90 percent of the new cases and deaths globally happened in low- and middle-income countries, where access to these interventions is more limited. But even within wealthier nations, disparities in cervical cancer remain—and they often fall on racial and ethnic lines.

Māori women are diagnosed with cervical cancer at more than twice the rate of New Zealand European women and are three times more likely to die from the disease. Similar inequalities have been recorded in neighboring Australia, where rates of diagnosis and death from cervical cancer are two and three times higher among Indigenous compared with non-Indigenous women. And while the World Health Organization has acknowledged that inclusion of Indigenous populations in cancer registries globally is insufficient, evidence from countries where such data are collected indicate that Indigenous peoples tend to experience a higher risk of cancers and have worse health outcomes than their non-Indigenous counterparts.[78]

Low uptake of screening is often highlighted as a contributing factor. In New Zealand, for instance, 34 percent of Māori women don't attend regular cervical screening, compared with 21 percent of New Zealand European women. But again—whether in the US, New Zealand, or elsewhere—closing these kinds of health gaps requires not only encouraging people from marginalized racial and ethnic groups to get screened, but actively identifying and eliminating barriers that are preventing people from accessing these potentially lifesaving health checks.[79]

A recently trialed intervention in New Zealand provides an example. A 2019 survey identified a desire for bodily autonomy, including whakamā (embarrassment, shyness, or reticence), as the most frequently cited barrier to cervical screening among Māori women, and found that the majority of survey respondents said they would be likely or very likely to do a self-administered HPV test at home and to attend follow-up if they were to receive a positive test result. When these findings were subsequently put to the test in a randomized controlled trial, cervical screening rates were found to be almost three times higher among Māori women who were given the option of doing the screening at home as an HPV self-test, compared to those offered the standard cervical smear. Following the results from the trial, in May 2021 New Zealand's health ministry announced NZ$53 million ($33 million) worth of funding toward the implementation of a wider at-home cervical screening program.[80]

In addition to screening, early recognition of cancer symptoms is also vital. Yet as a study from the UK demonstrates, racial and ethnic inequities loom over this stage of diagnosis too. The analysis of more than 126,000 cancer cases in England between 2006 and 2016 found that people of Black or Asian ethnicity had to wait longer than White people to be diagnosed after first presenting to a GP with cancer symptoms. Whereas the median waiting time among White people was 55 days, among people of Asian ethnicity it was 60 days and among Black people it was 61.[81]

As anyone who has direct or indirect experience with cancer knows, the earlier it is detected the better a person's chances of survival. Every day counts. That is why screening and early recognition of symptoms are both so important. But cancer is complicated and difficult. The cancer journey doesn't begin nor end at diagnosis—and neither do the associated racial and ethnic inequities. The authors of the UK study acknowledged this complexity in a 2022 research article summarizing their findings. They pointed out that the observed inequities in waiting times for diagnosis were "unlikely to be the sole explanation for ethnic variations in cancer outcomes," which have previously been documented in the UK—as in the US, New Zealand, and elsewhere.

Indeed, many factors contribute to cancer risk and outcomes. So far, we have seen how cancer begins with changes to DNA, which can result in dangerous, uncontrolled cell growth. These changes tend to accumulate as we get older, which is why the risk of cancer rises with age, but environmental exposures can accelerate them too. The fact that some of these environmental exposures are very well known, such as radiation or tobacco smoke, can sometimes give the illusion that they are completely avoidable.

But for many of us, avoiding exposure to cancer-causing substances, known as carcinogens, is difficult—if not impossible. What if they are in the air you breathe? What if they are in the water you drink? What if they are in the materials you use to do your job? As we will see in the coming chapters, inequities in societies globally often mean that these sorts of unavoidable environmental exposures disproportionately affect people belonging to marginalized racial and ethnic groups. Racism is harmful to health in all sorts of ways—and it isn't only our physical health that is suffering.[82]

In the summer of 2021, I was experiencing some personal challenges and I decided to try seeing a therapist. I was encouraged by my mother, who had recently switched careers to counseling from pharmacy. I had seen an ad on social media for one of those apps where you can find a counselor or therapist online and attend appointments via phone or video call, and I thought, why not give it a go? I answered a few multiple-choice questions in the app, and then I was presented with a long list of potential therapists to choose from.

An initial, introductory video call with my chosen therapist went well. We exchanged some small talk and she told me a bit about her background and how she usually works. I told her about myself, including about some past mental health struggles I had experienced, and explained that I wanted to try counseling to see if it could provide me with some tools to help navigate present and future challenging situations in my life. I left that introductory session feeling positive and I immediately scheduled another appointment for a few weeks later. It was during that second appointment that I sensed something wasn't quite right.

My therapist, a White woman who I would guess was in her forties or fifties, became noticeably uncomfortable when I started speaking about my experiences of racism. As the conversation continued, I felt the need to start playing down my own experiences in order to protect her feelings. That made it difficult for me to speak freely about what was happening in my life at the time, as racism was a central part of it. After reflecting and talking about the experience with my mother and my partner, I made the decision to stop working with that therapist.

I hadn't thought much of this experience until I met Alfiee Breland-Noble, a psychologist and the founder of the AAKOMA project, a nonprofit organization in the US focused on illuminating and reducing mental health disparities for diverse communities. Dr. Alfiee, as she is fondly referred to by those around her, founded the AAKOMA project back in 1999. As a professor doing mental health research, she became frustrated at the lack of funding available in the US for research focused on better addressing the mental health care needs of Black people and other people of color. So, she decided to channel her ideas and passion into a nonprofit. At first, "I was funding it out of pocket, paying for everything and doing everything myself," she said. But, according to Dr. Alfiee, the general attitude toward her work shifted significantly following the murder of George Floyd, an unarmed Black man, by White police officer Derek Chauvin in Minneapolis in 2020.

Floyd's murder sparked a wave of anti-racism protests in the US and around the world, including in Germany, where I had recently relocated with my partner. "Then it exploded," said Dr. Alfiee, because the public's heightened awareness of racism and its harms was accompanied by a sharp rise in donations. The AAKOMA project has been using that donated money to provide free therapy sessions and workshops for teenagers and young adults of color in the US, as well as raising awareness and providing education for young people and their families on the importance of mental health. This kind of work is sorely needed, because, just as with physical health, there are vast racial and ethnic inequalities when it comes to mental health too.

It turns out the difficulty I faced in discussing my experiences of racism with my therapist is not unique—neither in Germany nor elsewhere. A 2020 survey of more than eight hundred Black people living in Germany found that 62 percent of respondents agreed with the statement "In psychotherapy, my experiences of racism are not taken seriously and are questioned." More than a third said that this was something they experienced "often," "often," or "very often."[83]

Failure by mental health care providers to acknowledge and process the realities of living in a racist society has been linked to poorer perceptions and experiences of care among people from racial and ethnic minority backgrounds, for instance in the UK and US. These kinds of negative perceptions and experiences can in turn discourage people from accessing such services altogether, contributing to disparities in access to mental health support for people of color and people belonging to marginalized racial and ethnic groups globally. "It's not our identities impacting our ability to access care, it's the system's inability to recognize and understand and incorporate our needs," Dr. Alfiee pointed out to me, after I told her about my experience with the therapist. "You should never be in a position where you feel like you have to tone down your experience. That's not why you're there."[84]

I was fortunate enough to be in a position where I felt able to seek support in the first place. The whole subject of mental health is shrouded with stigma, especially when it comes to mental illness, and research from the US, the UK, and elsewhere suggests that such stigma tends to be higher among racial and ethnic minority compared with majority groups. Cost presents another major barrier to accessing mental health support, which Dr. Alfiee explained to me is often a particular issue for Black people and other people of color in the US, where she is based. "Then imagine, on top of that, you get past the hurdle of money, you get past the hurdle of finding, you know, getting access to some of this care. You get somebody, and then you do all of that and they're making you feel like you need to tone down who you are," she said. "Come on."[85]

A few months after my conversation with Dr. Alfiee, I found a different therapist, through my mother's network in the Netherlands, where she has her own counseling practice. That therapist, who I ended up doing multiple sessions with, is a White woman, and from the outset she was very open to talking about racism and other issues related to my identity and life experiences—including acknowledging and discussing these dynamics in the context of our therapist-client interaction. I felt really grateful to have found a therapist who could support me fully, in all aspects of my identity. Unfortunately, this kind of support just isn't accessible for many people belonging to marginalized racial and ethnic groups around the world—and it is even less so for people with multiple intersecting marginalized identities. Tragically, these are often the very people who could benefit from this type of support the most, given the harmful effects that experiencing racism and discrimination can have on mental health and well-being.[86]

"In the mental health space, we're a helping profession. You can't be a helping profession and not help the fullness and wholeness of what a person brings to the table. And so, when we talk about cultural competence, that's what we're talking about," said Dr. Alfiee. "We're not talking about; you need to know every single thing there is to know about every single culture. What you can do is be open enough and have a skill set, have a knowledge base and have an awareness level that allows you to be present and welcoming and inviting and supportive of anybody who comes into that psychotherapeutic space, seeking your help," she said. "We all deserve the right to go into that psychotherapeutic space and be seen 100 percent for who we are."

On top of disparities in access to mental health support, in many countries there are also racial and ethnic inequalities when it comes to the diagnosis and treatment of mental illness. In the US, for instance, research by Dr. Alfiee and colleagues has found that university students of color with mental health conditions are less likely to be receiving treatment compared with White students. Similar inequalities have been documented among children and adolescents. "We know from data that if you take two, let's stick with young people, with the same set of symptoms, where most of us might associate those symptoms with an illness like depression—if you take two young people, put them in front of a professional with those same symptoms, and one of those children happens to be White, the other happens to be Black, with the same symptoms, the White child is more likely to be referred for services that are behavioral health in nature, the Black child is more likely to be referred for services that are more punitive in nature," Dr. Alfiee told me. "That's some of the variance in why so many more Black children end up in the juvenile justice system, end up getting suspended from school, end up getting expelled from school," she said.[87]

Inequalities in the treatment of depression extend to the adult population too. According to data for New York City from the Mayor's Office of Community Mental Health, just 30 percent of Black New Yorkers with depression self-reported receiving mental health treatment in 2017, compared with 58 percent of White New Yorkers. Black, Latinx, and Asian American and Pacific Islander New Yorkers with depression were all less likely than White New Yorkers to self-report receiving mental health care. "Untreated depression, untreated anxiety, exposure to trauma—if those things are left untreated, people behave in ways that are not the healthiest," Dr. Alfiee explained. "The result of unhealthy behavior, often, for Black people is incarceration. The result of unhealthy behavior for White people is support and help. Not always but often," she added.[88]

There are similar inequalities in other countries, such as the UK. A 2021 study reported that the likelihood of receiving treatment among people with common mental health conditions in England—including depression, anxiety, panic disorder, obsessive-compulsive disorder, post-traumatic stress disorder, and social anxiety disorders—was lower among all ethnic minority groups than among White ethnic groups, and lowest among Black people. At the same time, evidence from the UK, the US, and several other countries, including Canada, the Netherlands, and New Zealand, suggests that Black people and people belonging to other minority ethnic groups are more likely than White people to be admitted to the hospital involuntarily for psychiatric care. A study that reviewed data on psychiatric hospitalization in the UK and internationally found that among the most common explanations provided for the higher rates of involuntary hospitalization among racial and ethnic minority groups were an increased prevalence of psychosis, an increased perceived risk of violence, and increased police contact. Separate research indicates that clinicians are more likely to put emphasis on psychotic rather than depressive symptoms when examining Black compared with White patients—something that may be linked to an overdiagnosis of schizophrenia and underdiagnosis of depression among Black people.[89]

Harmful stereotypes about Black people—and Black men in particular—being dangerous and prone to violence may also contribute to increased police involvement and involuntary hospital admission among those experiencing mental illness, with potentially devastating consequences. An analysis of police killings in the US between 2015 and 2019 found that unarmed Black men exhibiting signs of a mental illness were more likely to be shot and killed by police compared with

White men exhibiting similar behaviors. This comes on top of a backdrop of disproportionate police killing of Black people, which itself contributes to racial trauma and harms mental health. Underlying all of this inequality is a historical legacy of racism embedded within the field of psychiatry itself.[90]

"There was literally a White psychiatrist who made up a name of a psychiatric illness for enslaved Africans who tried to escape," Dr. Alfiee told me. The psychiatrist she was referring to was Samuel Cartwright, a doctor and slaveholder in the US who in the 1850s hypothesized that enslaved Africans fleeing captivity were doing so because they were experiencing a mental illness that he termed "drapetomania." "The idea was that if you, as a Black person, wanted to escape bondage, there was something wrong with you psychologically," said Dr. Alfiee.[91]

The fight by Black people in the twentieth-century US for civil rights was similarly pathologized by psychiatrists as a supposedly schizophrenia-like condition termed "the protest psychosis," as US psychiatrist and author Jonathan Metzl illustrates in his book by the same name. "You still have stereotypes, all through history, of Black people, I think in the United States and globally, being seen as everything from savages, to lazy, to ignorant," said Dr. Alfiee. "If you think about that history, and then if you think about the field of psychiatry overall, and you come all the way up to current day, people have long memories."[92]

The examples of racial inequity that we have covered so far, from Serena Williams's childbirth experience in the US to Saidy Brown's journey with HIV in South Africa, are by no means an exhaustive list. I have deliberately chosen not to dedicate too much of this book to simply documenting disparities. While documenting inequality is undeniably important, it doesn't go very far toward identifying underlying causes.

Now that we have established that health gaps exist, I want to start investigating *why* they exist. I want to examine how racism in society, medicine, and science contributes to health gaps and harms health more broadly. Of course, society, medicine, and science are overlapping spheres—medicine and science exist within society, for one thing—but for the purposes of our investigation we are going to focus on each of them in turn. I hope that in doing so—with help from patients, activists, medical students, doctors, and scientists, who we will meet along the way—we can illuminate potential solutions.

Systemic racism has been lurking behind everything we've discussed, but in the next chapter, I want to explore head-on how this insidious form of discrimination contributes toward the health gaps we have learned about so far. Because systemic racism influences every aspect of our lives, from the health care we can access to the very air we breathe.

PART II

Racism in Society

CHAPTER 3

Systemic Racism

The situation was desperate. People were dying at home for fear of going to the health centers, which were full of people sprawled on the floor, waiting for beds. "We saw entire families decimated by Covid," Marcos Antônio dos Santos Junior told me, recalling how the first wave of Brazil's Covid-19 epidemic in 2020 tore through his community in southeast Brazil.

Marcos lives in São João de Meriti, a city on the outskirts of Rio de Janeiro, which has earned the nickname "formigueiro das Américas" or "anthill of the Americas" because it has one of the highest population densities of any city in Latin America. Like most other cities of the Baixada Fluminense—part of the wider Rio de Janeiro metropolitan area—it is also known as a dormitory city, according to Marcos, with many inhabitants commuting to the center of Rio de Janeiro to work during the day and then returning to São João de Meriti to sleep at night. He told me that although he loves his home city, he recognizes that it has "great needs," particularly when it comes to health.

Brazil has a system of free, universal health care, but its public hospitals are badly underfunded. The situation is especially dire in low-income suburbs, favelas, and cities, including São João de Meriti, where populations are disproportionately "preto," meaning Black, and "pardo," meaning Brown or Mixed with African ancestry. According to Brazil's 2010 census, its most recent one at the time of writing, about 63 percent of the population of São João de Meriti self-identified as Black or Mixed, in comparison to 48 percent in neighboring Rio de Janeiro. Black and Mixed Brazilians are also more likely than White Brazilians to rely exclusively on the public health care system, because of economic barriers in accessing private health insurance. Marcos, who identifies as Black, pointed to this as one of several examples of systemic racism, which he believes is the main driving force behind differences in health along racial lines in Brazil.[93]

"There are very poor neighborhoods in São João, people who are below the poverty line and have no help from the public authorities," he explained. "Unfortunately, to have access to quality services, it is necessary to move from our city to the capital, and, even then, it is often necessary to have good financial conditions to achieve rights that should be offered by the public authorities," he said. "When I look around, I am aware that being Black here in Brazil is very difficult."

Marcos and I began corresponding in November 2021, with the help of online translation software (I don't speak much Portuguese and he doesn't speak much English). He told me about his family, before describing the pain and loss they endured when Covid-19 first reached their city the previous year. "I had the privilege of having a father who was a refrigeration mechanic, and the company that my father worked for provided health insurance, which contributed a lot to my health," he said. "I trained as a mechanic and also managed to work in a company, which provided health insurance, even though it was not my dream job," he added. But Marcos explained that being Black in Brazil feels like having the odds stacked against you. Even if a White person and a Black person are of equal social class, the Black person will certainly have more difficulty in achieving their goals, he told me. "With me and my family it was no different, we needed to fight a lot to have our basic needs met," he said.

When Covid-19 hit São João de Meriti, Marcos felt he had little choice but to continue commuting to work each day in crowded public transportation, all the while praying to God to keep him and his family safe. After all, his job was what allowed him to take care of his and his family's health. But a preexisting heart problem left his father, in his late sixties at the time, particularly vulnerable to the virus.[94]

"I lost my father to Covid," said Marcos. "It was very difficult for us, because it was not even possible to say goodbye and have a dignified funeral," he told me. At the same time, Covid-19 deaths in the wider community were mounting. "All the time we received news of death from neighbors and friends," remembered Marcos. Hospitals and health centers in the city were overwhelmed with patients, forcing many people to travel to Rio de Janeiro in search of urgently needed care.

In May 2020, a report by nonprofit investigative news agency *Pública* revealed that neighborhoods of Rio de Janeiro and São Paulo with majority Black residents were experiencing more Covid-19 deaths in comparison with Whiter neighborhoods. Subsequent research has confirmed that Black and Mixed Brazilians who were admitted to the hospital with Covid-19 during the first months of Brazil's epidemic lost their lives at a significantly higher rate than White Brazilians. "This pattern is explained by social inequalities and prejudice," said Marcos. "I think there is a difference in health between racial groups, due to institutional racism here in Brazil."[95]

Institutional racism, structural racism, and systemic racism are terms often used interchangeably to refer to the ways in which racism is deeply entrenched within the systems, structures, and institutions that underpin societies across the globe. More important than what we choose to label this form of oppression (I personally tend to call it systemic racism) are its devastating consequences on health.

Systemic racism is what is happening when Black Brazilians like Marcos encounter additional barriers to accessing quality health care that their White Brazilian counterparts don't. It is also what is happening when, as we've seen, maternity wards are disproportionately closed in Black neighborhoods in the US, or when Māori women in New Zealand face additional barriers in accessing culturally safe cervical cancer screening. Importantly, though, it is also what is happening *before* people reach the point of even needing to access health care. It is the added challenges that many people belonging to marginalized racial and ethnic groups globally face in attempting to lead healthy day-to-day lives—in accessing clean air to breathe, uncontaminated water to drink, nutritious foods to eat, and safe spaces in which to exercise.

To understand how systemic racism harms health, it is helpful to remind ourselves of its origins. In Brazil, as Marcos put it: "There is an almost unpayable debt from three hundred years of slavery."

Edna Maria de Araújo, a public health researcher at the State University of Feira de Santana in Brazil and a member of the Racism and Health Working Group from the Brazilian Association of Collective Health, agrees with Marcos and argues that to understand racial and ethnic inequalities in Brazil today, we need to consider the legacy left on the country by European colonization and transatlantic slavery. A 2020 research article by de Araújo and Kia Lilly Caldwell, a researcher in African, African American, and Diaspora studies at the University of North Carolina at Chapel Hill, among other academics, examined data showing vast racial and ethnic disparities in Covid-19 outcomes in both researchers' base countries—Brazil and the US—and pointed toward commonalities between the two nations that could explain these patterns of inequality. The researchers highlighted the fact that Black (and, in Brazil, Brown or Mixed) people in particular, the majority of whom are the descendants of enslaved Africans in both Brazil and the US, were also disproportionately impacted by the pandemic in both countries. Indeed, this chapter in human history contributed significantly to the systemic racism and health inequities that persist in many countries globally to this day.[96]

Of the estimated eleven million enslaved Africans transported to the Americas between the sixteenth and nineteenth centuries, it is thought that about five million were brought to Brazil: more than any other single country. The Portuguese were the first to make the transatlantic journey in the late 1400s, transporting enslaved humans bought from slave traders in West Africa across the ocean to Brazil. I remember learning about this history as a child during a visit to Ghana, where my maternal and paternal grandmothers were from and where my father grew up. My parents had taken my sister and me to visit Elmina Castle, a formidable structure erected on the Ghanaian coast by the Portuguese in the late fifteenth century, which became a major location for holding enslaved people captured from different parts of West Africa during the transatlantic trade. Our tour guide told us about the horrendous conditions in which people were kept there and the inhumane ways in which they were treated, before being packed into ships as cargo and destined to either die a horrific death at sea or to continue life in the Americas as someone's property. We learned also how Elmina Castle was taken over by the Dutch in the 1600s and then ceded to the British in the 1800s, both also major slave trading nations.

My sister and I could feel that we were the product of all of these histories: our parents are both half Ghanaian, we grew up in the Netherlands, we have British nationality, and we are both the descendants of enslaved Africans through our paternal grandfather, who was from Dominica, a small Caribbean island nation south of Guadeloupe. Our surname, Liverpool, suggests that our enslaved ancestor was transported from Africa to the Americas by the British via the port of Liverpool, which was a major slave trading port in the UK. It would take both of us a lot longer to fully appreciate the continuing legacy of these histories in our modern world.

As de Araújo and Caldwell illustrate in their research article, this legacy is very much connected to the disproportionate impact of Covid-19 on Black people observed in both Brazil and the US. Brazil became the last country in the western Hemisphere to abolish slavery in 1888, and although the country never had a legalized system of racism directly equivalent to the Jim Crow laws that enforced racial segregation in the US, employment discrimination and residential segregation have consistently limited opportunities for Black and Mixed Brazilians. This systemic racism continues to impact people today; the average income of Black and Mixed Brazilians remains about half that of White Brazilians, not entirely dissimilar from the picture in the US, where in 2019, Black men and women earned about 78 and 66 cents, respectively, for every dollar in average hourly wages earned by White men.[97]

Economic inequality like this also exists in many other countries around the world, including the UK, where people from British colonies and former colonies in Africa, the Caribbean, and Asia were encouraged by the government to immigrate in the mid-twentieth century to help boost the workforce in the aftermath of the Second World War. On arrival, these immigrants, among whom were my paternal grandparents, faced considerable racial discrimination, which was only made illegal in the UK in 1965—a year after my father was born in London. This made it more difficult for "colored" people to get jobs and to find accommodation, contributing to economic inequality: in 1964, a Conservative MP was elected in the constituency of Smethwick in England after endorsing the (uncensored) slogan "If you want a n★★★★★ for a neighbour, vote Labour." As in other countries, the legacy of this systemic racism can be felt in the UK to this day.

A report that examined rates of poverty in the UK prior to the Covid-19 pandemic found that 43 percent of Black households were in persistent poverty, compared with 19 percent of White households, for instance. And whether in the UK, the US, or Brazil, there is no doubt that being on low income is associated with worse health, including a higher prevalence of chronic health conditions that have been linked to poorer outcomes with Covid-19, such as cardiovascular disease and diabetes. Being on low income is also associated with reduced access to quality health care, particularly in the US, which is exceptional within the OECD—an association of mostly wealthy countries—in not having a system of universal health care. But as the experiences of Marcos and so many others show, even in Brazil, which has a universal health care system, similar inequities in access to care are present.[98]

"Those with low socioeconomic status are those who have limited access to the public health system and are unable to pay for private health care," de Araújo pointed out, recapitulating what Marcos told me. And in Brazil these people are disproportionately Black, she added. "We live in a country where racism is structural," said de Araújo. "I think it is necessary to deconstruct racism through the implementation of public policies that aim to repair iniquities committed against the Black population, not only in the area of health, but in all areas," she added.

Income inequality isn't the only way that systemic racism manifests in today's world, according to Darrick Hamilton, an economist at the New School in New York City. Hamilton is a prominent US policy adviser and a leading scholar in the field of stratification economics, which recognizes the importance of social hierarchies and structures in shaping people's economic outcomes. He told me: "A lot of our conversations around economic security and well-being have focused on income. But over the last decade or so, there's been a greater emphasis with regards to understanding perhaps a more paramount indicator of economic security: wealth."

As with income, wealth—which Hamilton defines succinctly as "the difference between what you own versus what you owe"—is a strong predictor of overall health and well-being. "If you're faced with some calamity that you didn't anticipate—you lose your job, there's a pandemic—it is that nest egg of wealth that allows you to endure," he said. Meanwhile, debt—the opposite of wealth—has been linked to higher perceived stress and depression, as well as poorer self-reported general health and higher blood pressure.[99]

Wealth, to an even greater extent than income, is unevenly distributed along racial and ethnic lines. In the US, for instance, the median Black or Latinx household has 10 or 12 cents in wealth, respectively, for every dollar owned by the median White household. This racial wealth gap, as Hamilton calls it, is a significant contributor to racial and ethnic disparities in health and may partly explain why racial health disparities often persist even after controlling for income.

"The racial wealth gap is an implicit indicator in economic terms of a long racist history in the United States that has really important contemporary implications," he told me. "If we go back to the beginnings of the nation, it was a period in which Black people served literally as an asset for a White, land-owning plantation class," he said. In the nation's more recent past, racist practices since outlawed—including redlining, a policy under which the Federal Housing Administration denied loans for housing in and near predominantly Black neighborhoods, and racial covenants, preventing non-White people from buying homes in certain neighborhoods—have deprived Black families of generational wealth.[100]

Hamilton suspects that this long-standing economic inequality contributes to worse health outcomes for Black people as well as other people of color, not only by fostering gaps in access to quality health care, but also by taking a direct toll on mind and body. "When we ask Black people to work twice as hard to overcome obstacles, that manifests in negative health. And what is pernicious and ironic, is that those who are perhaps social climbers in a relative sense, compared to White people, are predisposed to even greater health risks," he said. As an example, he pointed to the Black-White gap in the rate of infant mortality, which is present among infants with mothers at all levels of educational attainment, and which is even wider among those whose mothers are educated to master's degree level or higher. "When you find that a college-educated Black woman is more likely to suffer an infant death, if she's expecting, than a White woman who dropped out of high school," Hamilton continued, "then that's not only suggestive of a disparity across race, but one where, as Black people move up in socioeconomic status, the comparative disparities across race actually rise," he said.[101]

Writing this book forced me to rethink my understanding of how systemic racism harms health. Hamilton's argument and the evidence he cited made me realize that the relationship between economic inequality and health is far more complicated than it might first appear. Like many people, I had previously always assumed that if a person could overcome the odds and improve their economic situation in spite of all the barriers that exist within our unequal world—including those resulting from systemic racism—then that person would reap all the benefits in the form of better health. After all, higher socioeconomic status is clearly associated with improvements in health within all racial and ethnic groups. But this assumption ignores the fact that overcoming the odds—particularly overcoming systemic racism—can itself come at a cost to health.

"When we tell people to work twice as hard to get by, you're asking somebody to be above average. You're asking somebody to be super—something that is above a norm," said Hamilton. "Well, with any system, there's costs to that." This may partly explain why the health gains associated with increasing socioeconomic status—measured in Hamilton's earlier example as the level of educational attainment among Black or White mothers in the US—tend to be shallower for those most likely to be disadvantaged by systemic racism compared to those most likely to benefit from it.

There is another issue with my previous assumption: it accepts the premise that a person's health should depend on their economic situation and puts the onus on the individual to improve their situation and their health. This might work if we lived in a world where everyone had equal access to opportunities and resources, but that clearly isn't the case. And I think part of the reason I didn't fully appreciate this earlier is because of my own privileged position.

I am fortunate to have had a very privileged upbringing. My parents immigrated to Europe in the 1980s—my father from Ghana to the Netherlands, and my mother from Lebanon to the UK (my father was actually born in the UK, but his family moved back to Ghana when he was a baby and he grew up there). Both my parents went to university, and they worked hard to ensure that my sister and I had everything we needed and more so that we could achieve good grades in school and have access to more opportunities than they had. We were one of the only Black families in our neighborhood in the Hague, where I was born and raised, and my sister and I were one of few Black children at the elite, international school my parents forked out to send us to.

I think my parents both have what my sister and I now often jokingly refer to as "good immigrant syndrome." They faced considerable racism in their lives, and they were desperate for themselves—and us—to fit in, be accepted, and even thrive within the predominantly White environment we were living in. There was a sense that we had to be "above average," as Hamilton put it—we had to be twice as good and work twice as hard, as if to compensate for our Blackness. I think this had a big influence on me. I took school extremely seriously as a teenager and achieved top grades in all my subjects. I was extremely motivated to go to university in the UK, like my mother had, and I received offers from four of the five universities I applied to there to study biomedical science.

My experience attending an elite, predominantly White school turned out to be ideal training for my life at university, where I was often one of the only Black students at the lectures I attended. I graduated with a First Class bachelor's in Immunology and Infection from University College London, before moving to the University of Oxford, where I had been accepted into a highly competitive PhD program funded by the Wellcome Trust charitable foundation.

In order to fit into these various White, elite worlds, in both the Netherlands and the UK, I was forced to internalize negative ideas and stereotypes about Black people (I have since learned that this is a common experience among minoritized people in that position—journalist Linda Villarosa describes going through something similar in her book *Under the Skin*). To make sense of these racist notions I had absorbed, I had to believe that my family and I were somehow exceptions. We "weren't like other Black people," as many of the White people around us frequently reassured us. Even when I experienced racism myself or when I was directly faced with the products of systemic racism—such as the fact that just 1 percent of university professors in the UK are

Black—I convinced myself that these were individual failures rather than systemic ones.[102]

The same faulty logic is often applied to racial and ethnic disparities in health, for instance with my earlier assumption that climbing the economic ladder against the odds would be sufficient to rescue Black people and other systemically marginalized people from poor health. This fails to recognize that our societies are fundamentally unequal in so many ways.

Covid-19 forced a greater reckoning with this inequality because it exposed the systemic nature of it so brutally. Consider Marcos in Brazil. At the height of the pandemic, Black and Mixed Brazilians like him were more exposed to the virus because they were less likely to be employed in jobs that could be done remotely and more likely to be commuting to work by public transport, like he was. The situation was similar in other countries including the US and UK, where Black people and other people of color were more likely to be working in jobs that were deemed essential during the pandemic and that couldn't be done from home. Ironically, many of these so-called essential jobs were poorly compensated: a 2021 analysis in the US found that essential workers constituted almost half of all workers in lower-paid occupations. Being on a lower income is associated with poorer health and increased vulnerability to diseases like Covid-19, so this is a vicious circle. And—as we are about to discover—economic inequality isn't the only way in which systemic racism manifests and harms health.[103]

The first time Rosamund Kissi-Debrah noticed something was wrong with her daughter, Ella, was in October of 2010. Ella was seven at the time and it was half term, a school holiday. "We were doing the Great Fire of London at school and we went to see the Monument," Rosamund told the BBC during an interview about a decade later. "Ella had a cold and she was climbing up the stairs. I remember her voice saying to me, 'I can no longer climb,' and me saying, as mothers do, 'You've only got a cold, what's stopping you?' I still feel really bad about that," she said. Ella managed to make it to the top, but she was exhausted during the train ride home and, uncharacteristically, fell into a deep sleep. Soon afterward she developed a strange-sounding cough, like that of a smoker, and within several weeks she had become so unwell that she had to be put into a medically induced coma.[104]

Over the next three years, Ella would be hospitalized almost thirty times after experiencing severe attacks of what was eventually diagnosed as a rare and complex form of asthma. "You know she was always fighting to breathe and it's something I will never get over. Seeing your child choke—that's how I call it—constantly, she used to collapse and stop breathing and then I'd have to resuscitate her and then she'd come back," Rosamund recalled, during an interview in a 2020 BBC documentary. "She said to me, 'I think I'm going to die,' and my insides, literally, you know it's like a vomiting feeling." In February 2013, Ella did die, following a severe asthma attack. "I can't believe how much she battled, but she suffered horrendously," Rosamund said. "She was nine."

In December 2020, following years of campaigning by Rosamund and others, a London court ruled that polluted air made a material contribution to Ella's death, making her the first person in the UK to have air pollution listed as part of their cause of death. Ella and her family lived very close to the South Circular Road, one of the busiest highways in London. Her school was about a thirty-minute walk away, mostly along a pavement beside traffic. Crucial evidence in Ella's case was a 2018 report by asthma and air pollution expert Stephen Holgate at the University of Southampton, which concluded that exposure to illegal levels of air pollution was a key driver of her asthma. Holgate had examined tissue samples taken from Ella's body, as well as data from pollution sensors located close to her home near the South Circular Road. The official cause of Ella's death was ruled as "died of asthma contributed to by exposure to excessive air pollution." Coroner Philip Barlow added that inaction by authorities to reduce levels of nitrogen dioxide and a lack of information given to Ella's mother both "possibly contributed to her death."[105]

Following the verdict, Rosamund told my former colleague and mentor, journalist Adam Vaughan at *New Scientist*: "This is my daughter, this is what happened to her, and we have proved it." During one of several conversations we had a year or so later, I asked Rosamund what she had been feeling in that moment. "I think relief, yes, sadness, probably sadness first. Also, the fact that she wasn't here, because she always wanted to know what was wrong with her," she trailed off. "It wasn't going to bring her back," said Rosamund. "But, for me, it gave me a sense of peace."[106]

Rosamund is a remarkable human being—I have been in awe of her since the first time we spoke. Beyond the ruling on the cause of Ella's death, she remains a tireless campaigner for clean air in London and beyond. She is extremely kind and patient, and she is absolutely determined to make the world a better place for her children—Ella's twin siblings—and for future generations. "You need to strive for change, you need to strive for things being better. That's why I do what I do," she told me. "I do hope that it does impact the next generation of young people coming up, that my children who are in their teens will be able to have better air one day." She is also extremely resilient. "You've caught me on a good day. I am ever hopeful," she told me. "There are parents who've lost children, and they die very soon afterwards. It changes your whole life," she said. "I am blessed. I have twins. And, yes, I have difficult days. But I do want to be here for them as long as possible."

It is obvious that Rosamund loves her children dearly and her face lights up whenever she shares memories about Ella. During one of our conversations, I asked her how she thinks Ella would want to be remembered. "That's really easy," she replied. "She would want people to know how kind she was, how much she loved her friends and her family," Rosamund told me. Ella was an "all-rounder," she said. She was very good in school and continued with schoolwork sent by her teachers even while she was in the hospital. She was also talented at sports, said Rosamund. "Not being able to breathe was the biggest challenge for her. But most things she found quite easy," she explained. "She was very good at cycling, she was very good at swimming, she was very good at football," said Rosamund. "But definitely, she considered all the other things she was doing as passing time—her ultimate ambition was to be a pilot." Rosamund's eyes brightened as she recalled taking Ella and her siblings on a trip to Bournemouth to see the Red Arrows—the UK's Royal Air Force aerobatics display team—a few years before Ella passed away. "At the time I thought, oh Lord, we're from Lewisham, in Bournemouth, and we were like practically the only Black family there on the beach, as you can imagine," she said, and we both laughed.

"Because of her obsession, we'd left home really early to come to Bournemouth, so she can watch the Red Arrows for bloody ten minutes," Rosamund said with a chuckle. "It became an annual thing, to go and see the Red Arrows." Ella absolutely loved it; she was determined to get her asthma under control so that her health wouldn't interfere with her dream of one day becoming a pilot. "They've dedicated an airfield and a plane to her—and it's to encourage children from backgrounds like her to become things like pilots," Rosamund told me proudly. "If she was here, she would really like that."

Ella isn't here, though—and her tragic experience is unfortunately part of a much wider problem. One UK analysis found that Black children and adults were about twice as likely to be admitted to the hospital for asthma as their White counterparts, while the risk among South Asian children and adults was almost three times higher. In London, where Ella lived, a 2016 study for the mayor found that people from Black ethnic groups accounted for 15 percent of all Londoners exposed to nitrogen dioxide levels that breach European Union limits, despite making up just 13 percent of the city's population at the time. Ella's mother, Rosamund, and I spoke about these inequalities during a conversation the day after what would have been Ella's eighteenth birthday.[107]

"Lewisham has one of the highest asthma rates in London," said Rosamund. She and Ella's siblings still live in the same house in Lewisham in south-east London. "That is one of the criticisms of me, that why don't I move?" she told me. "Number one, I can't afford it—but that's not even the point," said Rosamund. "Also, when you move to these more exclusive areas to do with clean air, the [ethnic] diversity is less," she pointed out. "Look, if I want to go and buy plantain and yam, I just go down the road. I take it for granted. If I want to go and buy kenkey [a West African dish] or anything, I just go down to Catford. I don't think people get that, but anyway. And also, my friends are here—my friends who have supported me through this whole thing, they are here," she said. I could very much relate, especially as a child of Ghanaian immigrant parents and as a general plantain lover, but also—more importantly—simply as a human being: the suggestion that people should leave their homes and communities to escape from dirty air is not only nonsensical (where will we all go, once we've polluted all the air?), it is also racist.

Black people and other people of color are disproportionately exposed to air pollution, not just in London, but across England. A 2022 analysis by Friends of the Earth, an international environmental campaigning network, found that people of color in England are three times more likely to live in a very highly polluted area than White people. Very highly polluted areas are also more likely to be deprived, suggesting that those who are the most likely to be exposed to the highest levels of pollutants are also the least likely to be able to afford to move elsewhere.[108]

"In order to be able to pick and choose where you live, you need to be earning and how many of us are in that situation?" Rosamund pointed out during our discussion. "There is a [less polluted] area near-ish, if we had money, I'd like to move to. It's way above our budget—way, way, way above—but it's not too far from here. But we are not in that income bracket. God, hell no, we're not," she said.

The inequities in air pollution exposure that cost Rosamund and her family so dearly in the UK are mirrored on the other side of the Atlantic, as well as in many other countries around the world. "Environmental racism isn't something that has been linked to the UK much, but it is here," said Rosamund. "When I looked into environmental racism in the States, oh my god, it's even worse," she told me.

Indeed, a 2019 study in the US found that Black and Hispanic people were on average more exposed to air polluted with PM2.5—particulate matter less than 0.0025 millimeters across—compared to White people, despite generally contributing less to air pollution. The prevalence of asthma is also higher among Black people in the US compared to White people. According to the US Office of Minority Health, non-Hispanic Black people died from asthma-related causes at nearly three times the rate of non-Hispanic White people in 2020. The disparity is larger for children, among whom asthma is generally more common than among adults: non-Hispanic Black children were about five times more likely to be admitted to the hospital for asthma compared to non-Hispanic White children in 2019, for instance.[109]

Climate change is exacerbating these sorts of racial and ethnic health gaps, according to Yoshira Ornelas Van Horne, an environmental health researcher and justice advocate who was previously based at the University of Southern California but has since moved to Columbia University. Ornelas Van Horne's research is focused on addressing unequal exposures to harmful contaminants that affect structurally marginalized communities—something she understands intimately. "I grew up in a community, you know, Phoenix [Arizona], west side Phoenix, and it's kind of one of those communities that is labeled by academia as being an environmental justice community. So, one that lacks access to trees, green spaces. It's now divided by a freeway, which, in terms of asthma, living near freeways is one of those things that contributes to worse respiratory health outcomes," she told me. "For me, I think just seeing that community and other ones that I could relate to having to bear the brunt of environmental contamination is what really drove me to pursue this research," she said.

I reached out to Ornelas Van Horne after hearing her speak about her wide-ranging research on a podcast produced by Agents for Change in Environmental Justice, an organization at which she serves as assistant director for curriculum development. She told me about research she had been involved in, examining the health effects of "playa dust"—wind-blown particulate matter—in an area of southern California. The area, Imperial Valley, has a majority Hispanic and predominantly Mexican American population, and is located near a salt lake called the Salton Sea. The lake was formed in 1905, when heavy rain and snowmelt caused the nearby Colorado River to swell and flood the area. It has since been sustained by runoff from farms close by. But this source of sustenance is shrinking due to growing water scarcity—and so is the Salton Sea.[110]

"What we've seen is that this lake has been drying up—and that's led to the uncovering of what we call the playa dust," Ornelas Van Horne told me. "Playa is the Spanish word for beach," she clarified. "There's a lot of dust that occurs in this area, because it's also a very desert-prone area, so what we've seen is that there's been a lot of community anecdotes that they're not able to go outside when these high winds are going on, because it's affecting the health of them and their children," she said.

To examine these health effects more closely, Ornelas Van Horne and her colleagues surveyed 456 parents of elementary school children in the area about their children's symptoms and compared this with the children's estimated exposure to particulate matter air pollution using data from regional environmental monitors. "We've actually found that in this community, they have about 22 percent prevalence of asthma, which is extremely high," she said. "It's one of the highest in California." These findings are consistent with earlier research, which has shown that rates of asthma-related emergency visits among children in Imperial Valley are double the state average. In their preliminary study, Ornelas Van Horne and her colleagues also found an association between increases in particulate matter air pollution in the environment and increases in reports of respiratory symptoms, such as wheezing, as well as use of asthma medication among the children. "Really our hunch is that there's something in the immediate environment that's contributing to their symptoms," she told me. "And so, we're really trying to investigate, what is it about this dust? I'm particularly looking at pesticides," she said.[111]

Replenishment of the Salton Sea over the years with runoff from nearby farms means the lake bed contains lots of pesticides, in addition to other potentially toxic substances in the sediment, including metals such as lead and arsenic—all of which can become mobilized as dust when wind blows over dried-up areas of the lake bed. Globally, environmental injustices like this are likely to become more pronounced as the planet warms, warned Ornelas Van Horne.[112]

As climate change drives changes in land use and diminishes water resources, the production of wind-blown dust and dust storms is expected to accelerate. India, which experiences regular dust storms, is already seeing rises in the severity and frequency of these extreme weather events. In 2018, northern India was hit by three severe dust storms in a row, which were found to be associated with increases in particulate matter concentrations. Air pollution levels in India are already among the highest in the world and there are also vast inequalities in exposure within the country. Districts with higher percentages of people belonging to traditionally marginalized caste groups have been found to have higher average PM2.5 concentrations, for instance.[113]

Environmental racism is everywhere—another insidious manifestation of the systemic racism that harms the health of people belonging to marginalized racial and ethnic groups around the world. This is not a new phenomenon. In fact the term "environmental racism" originated in the US in the late twentieth century, coined by civil rights leader Benjamin Chavis. A landmark report published in 1987 by the United Church of Christ Commission for Racial Justice revealed that race was the most significant predictor of a person living near hazardous waste, and in 1990, sociologist Robert Bullard's book *Dumping in Dixie* linked the siting of hazardous waste facilities with historical patterns of segregation in the southern US. The report and Bullard's book are often cited as part of the beginning of the country's environmental justice movement and, unfortunately, their findings are just as relevant today as they were when they were first published; more recent analyses have identified similar associations between race and the location of hazardous waste sites in the US.[114]

In addition to pollutants in the air people breathe, there are also racial and ethnic inequities when it comes to exposure to other environmental pollutants and contaminants. A 2019 report showed that communities of color in the US experience higher than average rates of drinking water violations, where drinking water fails to meet legal safety standards. This disparity is exemplified by severe water crises in Flint, Michigan, and Jackson, Mississippi, two cities with majority Black populations.[115]

In Flint, for example, residents were exposed to lead and other contaminants in drinking water after the city switched its water supply from Detroit-supplied Lake Huron water to drawing water from the Flint River in 2014. The effect on residents' health was visible in a study conducted around the same time; it found that within the Flint area, the incidence of elevated blood lead levels among children under five increased following the switch. It is worth pointing out here that there is no safe level of lead exposure and that young children are particularly vulnerable to lead poisoning, which can cause permanent damage to the brain and central nervous system. Poor Black children are especially at risk; one study found that across the US, Black children living below the poverty line were four times as likely to have elevated levels of lead in their blood compared to White children living in poverty.[116]

But all this just begins to scrape the surface of the relationship between race, environment, and health. More than six hundred miles south of Flint there is a region along the Mississippi River in Louisiana that houses more than a hundred oil refineries, plastics plants, and chemical facilities; it has earned the nickname "Cancer Alley" as a result of the health issues experienced by its mainly Black residents. In 2021, a team of independent UN human rights experts described the situation as "environmental racism" and called for plans to further industrialize the region to be called off.[117]

To take another example: historical hard-rock mining in the western part of the US left a legacy of more than 160,000 abandoned mines, which are, for the most part, on the lands of Indigenous peoples—and there is evidence that Native American populations living near abandoned uranium mines experience elevated risks of kidney disease and hypertension. Ornelas Van Horne was part of a team of researchers who investigated the health impacts associated with the accidental discharge in 2015 of three million gallons of acid mine drainage from the Gold King Mine near Silverton, Colorado, into Cement Creek, a tributary to the Animas and San Juan rivers. The nasty-looking, yellowish-brown liquid began spilling out on August 5 when a crew contracted by the Environmental Protection Agency to inspect the mine inadvertently disturbed a layer of rock and dirt that had been sealing its opening. A risk assessment conducted by the EPA, which considered a recreational scenario for hikers, concluded that continuous exposure to sediments through a daily water intake of a half gallon per day from the river would be unlikely to cause adverse effects over an extended time period. But according to Ornelas Van Horne, this risk assessment failed to consider the intimate interaction that local Indigenous populations, including the Diné or Navajo people, have with the San Juan River.[118]

"The community were really the ones that helped us formulate our research, because they were like, 'Well, I use this river for way more things than just hiking, like, my children, we pray with this water, we irrigate with this water,'" Ornelas Van Horne told me. To address this oversight, in the summer of 2016 she teamed up with local community members and researchers at the Diné Environmental Institute, the Navajo Nation Department of Health, and the University of Arizona to conduct several focus groups, as well as a survey of 63 adults and 27 children living in three Navajo communities along the river. Together they identified 43 unique activities between the Diné and San Juan River and demonstrated that since the Gold King Mine Spill, the average number of activities each person in the survey reported engaging in had fallen by 56 percent. This was associated with considerable trauma for community members.

"A lot of them brought up these themes of historical trauma—this isn't the first time that something like this has happened; there had been a previous river spill decades before that nobody came to help them with, their community has been impacted by the mining industry for decades. Not to mention, they consider really the first trauma to be that of colonization," said Ornelas Van Horne. "They perceive the spill to be extremely detrimental to their health and to cause a lot of anxiety and worries about the future, because a lot of them didn't know how they were going to explain this to their children, or if they were going to be able to go back to doing their cultural activities, which they consider to be protective factors toward their health," she added.

Taking into account all of the activities between the Diné and San Juan River, Ornelas Van Horne and her colleagues worked with the affected communities to develop a community-based risk assessment and to implement an environmental sampling protocol. "Overall, we did see that the levels of at least two contaminants that we were mainly concerned with, which were arsenic and lead, were below the [limits] that the Navajo Nation had for some of these activities," Ornelas Van Horne told me. "This brings up a really good point of having communities set their own standards and be in charge of their own data," she said. "What we advocate for the community is really building infrastructure, not just for the Navajo Nation, but other communities, to be able to do and have continued environmental monitoring." This is important, as long-term uncertainties remain about the accumulation of contaminants in the San Juan River—as well as about contamination from the thousands of other abandoned mines in the region and in other parts of the country. "There's over ten thousand abandoned mines in that area alone," Ornelas Van Horne pointed out. "We're not really able to know the status of the other ones or how much contamination they may be polluting into that river or other ones, unless we're actively monitoring for contaminants," she said.

Similar instances of environmental racism can be found across the border in Canada. After a 2020 study found an association between long-term mercury exposure and premature mortality among people of the Grassy Narrows First Nation community, lawmakers voted to collect more data on the impact of environmental contaminants on racial and ethnic minority communities. The mercury was released into the community's aquatic ecosystem from a chemical plant in the 1960s in what has been described as one of the worst environmental disasters in Canadian history. Meanwhile, in Ecuador, thousands of Indigenous peoples are still living with the health effects of contamination from an oil spill in April 2020, which contaminated the Coca and Napo rivers—key sources of water and food for local communities.[119]

Environmental racism and inequity are also evident on a global scale. According to the WHO, outdoor air pollution caused an estimated three million premature deaths worldwide in 2012, with 88 percent of these deaths occurring in low- and middle-income countries, and the greatest number in the WHO Western Pacific and South-East Asia regions. And it is no secret that countries in the global south, including large populations of color, are more likely to experience the impacts of climate change—including health effects.[120]

The environments we live in have an enormous impact on our ability to live healthy lives. This fact was glaringly obvious to Marcos, when he reflected again on the challenges facing his and other communities on the periphery of Rio de Janeiro in Brazil. "We need basic sanitation," he told me. "Many communities and peripheral neighborhoods do not have access to this." Poor sanitation is linked to poor health through increased transmission of infections. A 2021 study in Brazil suggested a possible relationship between poor basic sanitation and Covid-19 cases, for instance. That study didn't include analysis of data on race or ethnicity but research from other countries strongly suggests that inequalities in people's living environments contributed to racial and ethnic disparities in disease and death during the pandemic. In the UK, household overcrowding was associated with increased Covid-19 risk among ethnic minority groups, for example. And studies in the UK and US have pointed to increased air pollution exposure as a contributor to more severe Covid-19 and more deaths from the disease among marginalized racial and ethnic groups.[121]

The more we unravel the economic and environmental impacts of systemic racism across societies, the less surprising racial and ethnic health gaps become. We have already examined racial and ethnic disparities associated with several illnesses, including the world's biggest killer—cardiovascular disease—a person's risk of which is strongly influenced by their diet and physical activity level. But overlapping economic and environmental inequalities influenced by systemic racism mean that not everyone has equal access to an environment with affordable, healthy food, or to safe, green space with clean air in which they can exercise.

In several countries, including the UK, the US, and Brazil, there's evidence that people belonging to marginalized racial and ethnic groups and people on low incomes are more likely to live in so-called food deserts, areas with little or no access to healthy food, or in food swamps, where unhealthy food options dominate over healthy ones. Living in unhealthy food environments has been linked to obesity, which is a major risk factor for cardiovascular disease among other health conditions—and there's evidence that rates of obesity are higher than average among Black and Hispanic people in the US and among Black people in the UK, for instance.[122]

There are similar inequalities when it comes to access to green spaces, which provide places for people to exercise or to just spend time in nature, with vast physical and mental health benefits. I have been an avid jogger since living in Oxford, in southeast England, where I was lucky to have access to lots of beautiful running routes in my immediate surroundings. Yet across England, there's evidence that areas with high proportions of Black and other ethnic minority residents tend to have fewer green spaces, compared with Whiter areas. According to a 2010 report by the UK's Commission for Architecture and the Built Environment, since merged into the Design Council, areas where fewer than 2 percent of residents are "Black and minority ethnic" have six times as many parks on average than areas where ethnic minority groups account for more than 40 percent of residents.[123]

In countries, including the UK, the US, and Brazil, where Black people (and Black men in particular) are often overly policed and more likely to be the victims of violent crime, concern about safety is another factor that can discourage people from exercising outdoors. The murder of Ahmaud Arbery, a twenty-five-year-old Black jogger, in a racially motivated hate crime in 2020 illustrates why. Ahmaud was followed by three White men—father and son, Gregory and Travis McMichael, and their neighbor, William Bryan—before being cornered and fatally shot, as he jogged through a predominantly White neighborhood in Georgia. A 2017 study in the US found that middle-class Black men living in majority White neighborhoods tend to exercise at lower rates than those who live in majority Black or racially diverse neighborhoods. This is despite the fact that majority White neighborhoods tend to have more facilities for leisure-time physical activity and more green and walkable spaces in comparison with Black neighborhoods, the study notes. Marcos highlighted similar concerns around safety in his neighborhood in São João de Meriti, Brazil, during our correspondence. He mentioned that children in his neighborhood don't feel completely free when they play soccer outside, because of fears about violence. Research in Brazil has found that Black people are more likely to be the victims of homicide, in comparison with people belonging to other racial and ethnic groups.[124]

Marcos and his wife, Élida, have taken an active role in supporting members of their community to participate in sports and eat healthy diets. In 2014, they launched "Projeto Inclusão"' or the Inclusion Project, to provide educational support to local teenagers having difficulties in school; during the pandemic, they expanded the project's outputs, for instance, helping to provide food to struggling families in their community. The project has since grown even further. "Today in the project we conduct sports activities, education, arts, vocational courses and encourage preventive medicine through healthy eating, because we know our reality, so we encourage the population that taking care of health is better than taking care of disease," Marcos told me. While efforts like the Inclusion Project are extremely commendable, on their own they cannot undo the centuries of systemic racism and inequality that have contributed to worse health for Black people and other people of color in many parts of the world. Governments globally must do more to ensure everyone has access to the resources needed to live a healthy life, by addressing systemic racism in all its forms.

The story doesn't end here though, because in addition to the systemic forms of racism in societies we discussed so far, other forms of racism—from chronic daily discrimination to acute racial trauma—also contribute to racial and ethnic inequities in health and well-being. So I decided to delve into the latest research examining the health harms of these more interpersonal forms of racism—and ask whether and how these harms might be transmitted intergenerationally.

CHAPTER 4

Interpersonal Racism

When his flight touched down in Charlotte, North Carolina, in 2022, Clint Smith was aware of the restlessness among his fellow passengers. Charlotte is a major travel hub, so many people had connecting flights they were anxious to catch. As soon as the fasten seatbelt sign switched off, several people stood up and began jostling to retrieve their bags from the overhead compartments before attempting to squeeze through the aisle to the front of the plane to disembark. Amid the jostling, one interaction began to attract attention. Two passengers, a Black woman and a White woman, both middle-aged, had bumped into each other during the rush and started arguing. Clint was standing close by. As they got off the plane, the White woman turned to the Black woman and—red with anger—called her the N-word. Clint locked eyes with the Black woman, as the White woman disappeared off into the crowded airport terminal.

Clint happens to be a poet and an author, who has written extensively about Black history, racism, and slavery in the US. He decided to write about how the altercation at the airport had affected him, as a Black man. He posted about it on social media, expanding on the experience in an article for *The Atlantic* a few weeks after the incident.[125]

We can only imagine how that Black woman must have felt in that moment and how the experience affected her. But what struck me about Clint's account was the extent to which the incident had also affected *him*, both mentally and physically, even though he hadn't been the direct target of the racist abuse. "It was as if my skin was struck by a match and fire spread through my entire body. My heart's once metronomic tempo accelerated into a gallop, my blood pumping as if it was trying to tell me to run away," he observed in the *Atlantic*. Clint noticed that for hours afterward, he felt the impact of the White woman's words in his body. "I couldn't shake it," he wrote. "Although the venom of her voice had not been oriented directly at me, I experienced the debris of her language. I felt it, quite literally, in and under my skin."

In addition to the systemic forms of racism that we explored in the previous chapter, interpersonal racism—in Clint's words, "intimate, direct, one-on-one racism"—can also affect a person's body and mind.

Experiencing racism in all its forms is exhausting, and as Clint astutely observed in his article, you don't have to be the target of an individual racist act to experience its harmful impact. There is burgeoning evidence that experiencing racism—or even just the anticipation of experiencing racism—harms people's health over time. Public health researcher Arline Geronimus first started thinking about this when she was a student at Princeton University in the late 1970s. Alongside her studies, she worked part-time at a school for pregnant teenagers in Trenton, New Jersey. Geronimus noticed that the teenagers were experiencing chronic health conditions that her wealthier, largely White Princeton classmates rarely did. She started to wonder whether there was a connection between the chronic health problems that these teenagers were experiencing and the stresses of their environment. Later, as a graduate student, Geronimus began her research by trying to address a question that we touched on in the first chapter of this book: Why do Black women in the US experience worse childbirth outcomes on average compared with White women?

During this time, in the 1980s and early 1990s, there was widespread recognition that the twenties through early thirties constituted prime childbearing ages, with lower risks associated with childbirth compared to younger or older women. Geronimus observed that this wasn't a universal truth. "In particular for Black Americans the risk of poor birth outcomes increased with age from the mid-teens just straight on up, so that ages we sort of assumed were perfectly healthy childbearing ages were already higher risk for Black moms," Geronimus explained during an April 2020 interview on an episode of WNYC podcast *The United States of Anxiety*. "It led me to wonder," Geronimus continued. "Is there something happening in the lives of Black women that leads them to poor health earlier than White women?"[126]

As a possible explanation for the disparity, Geronimus proposed the "weathering" hypothesis in a highly influential paper published in 1992, by which point she was working as an assistant professor at the University of Michigan. It was a radical idea at the time—she proposed that the mental and physical stress caused by experiencing racism day in and day out could be damaging Black people's health directly, resulting in increased vulnerability to disease and death. "Because of the multiple chronic and often toxic stressors across your life course, you would experience a stress mediated wear and tear on your organs and body systems and cells that in effect leaves you more vulnerable to poor health—it causes a kind of accelerated biological aging and premature death," she said. "I called it weathering because of the two different meanings that that word has. You can talk about weathering as being exposed to things that erode you, like 'the rock was weathered by the storm,' and you can also talk about weathering as in coming through a storm, as in 'the business weathered the recession.'"[127]

As scientific understanding in the fields of stress physiology and epigenetics grew over the next two decades, evidence for the weathering hypothesis began to mount, improving our understanding of how racism impacts physical and mental health. Researchers are increasingly discovering how the health effects of chronic exposure to discrimination within societies could help explain racial and ethnic disparities in health and disease outcomes that aren't accounted for by other factors. Living in a racist society can harm the health of all Black people, even those who don't directly experience racism, according to Delan

Devakumar, a public health researcher at University College London. "This is akin to other environmental risk factors for health, such as high levels of air pollution," Devakumar told the *Economist* in a 2020 interview. Early evidence for this phenomenon began to emerge in the 1990s. At the time, separate, parallel research was beginning to reveal the impact that the stress of experiencing racism has on the body.[128]

Several studies conducted in the 1990s and early 2000s demonstrated that exposure to racist provocation in a laboratory setting was associated with increases in heart rate and blood pressure levels among African American people. But racism isn't a one-off experience in a lab—and neither is the stress associated with it. The physiological consequences of chronic exposure to stress—including heightened levels of stress hormones like adrenaline and cortisol, as well as increased blood pressure—were well established by this point. The term "allostatic load" had been coined to describe the cumulative wear and tear that chronic exposure to stress responses has on the body over time. Allostatic load-scoring was established as a way of quantifying these physiological effects by combining measures of blood pressure and levels of stress-related biomarkers in the blood. Research by Geronimus and others in the US showed that Black people had higher allostatic load scores on average compared to White people, even after adjusting for other sociodemographic characteristics. Then, in 2010, another study led by Geronimus hinted that Black people might be experiencing accelerated biological aging compared to White people, as a product of greater allostatic load.[129]

A year earlier, the 2009 Nobel Prize in Physiology or Medicine had been awarded for the discovery of how chromosomes are protected by telomeres—caps of repetitive DNA at their ends, which shorten each time a cell divides. Telomere length can therefore provide an indicator of how young or old a person is at the cellular level. In their study, Geronimus and her team found that Black women between the ages of forty-nine and fifty-five had shorter telomeres compared to White women of the same age, equating to an estimated cellular or biological age gap of almost eight years.

Research by Geronimus and others caught the attention of David Williams, a social scientist and public health researcher at Harvard University's School of Public Health. Williams was shocked by the gaps in health and life expectancy between Black and White people in the US. In a widely cited 2012 paper, he highlighted the fact that Black people had an overall death rate that was 30 percent higher than that among White people in 2007, and that Black people had higher death rates than White people for 10 of the 15 leading causes of death at the time, including for heart disease, cancer, and stroke.[130]

This Black-White life expectancy gap still exists. As of 2017, the average life expectancy at birth for a non-Hispanic White person in the US was 78.5 years, compared to 74.9 years for a non-Hispanic Black person. This gap is predicted to widen because of the disproportionate impact of Covid-19 on Black populations. As a public health researcher, Williams was interested in further investigating the contribution of racism to health disparities, but he felt that a method for measuring racism directly was lacking. "We measure self-esteem," he said in his now famous 2016 TED talk. "There's no reason we can't measure racism if we put our minds to it." Williams devised the Everyday Discrimination Scale to capture, as he has described it, "ways in which the dignity and respect of people who society does not value is chipped away on a daily basis." The scale has become one of the most widely used measures to assess perceived discrimination in health research. And work by Williams and others has since demonstrated that high levels of perceived discrimination among racial and ethnic minority groups are associated with an elevated risk of a broad range of illnesses, from cardiovascular disease to breast cancer to mental health conditions.[131]

Even just *the anticipation* of experiencing racism may be harmful to health. Shawn Utsey, a psychologist at Virginia Commonwealth University and an expert on racism-related stress, explained this concept to me. Stress isn't only caused by overt experiences of racism, Utsey said.

Often it is caused by the broader challenges associated with negotiating a racist society, he said.[132]

"Black folks don't necessarily have to experience racism to be stressed by racism. Just the anticipation that you will experience racism is stressful," Utsey explained. He gave me an example, drawn from his own lived experience as a Black man in Virginia. "If I'm about to take my family shopping at the mall in the White part of town, which is probably what we would do, because they would be better resourced," he began, "almost immediately, when planning that trip, I would have to factor in the possibility that we would be treated poorly because of our race." Utsey went on, "And I would experience stress, even before anything ever happened." However, he acknowledged, "it may never happen." "But the possibility—the reality of the physiological nature of stress will still visit me, because of that cognitive process that tells my body that I'm about to experience fight or flight," he said.

Listening to Utsey, I started to reflect on some of my own everyday experiences, which up to this point I had thought to be quite trivial and mundane—things like being followed around shops by security, having my hair touched by people without asking or being racially profiled at airports. I also thought back to less frequent but perhaps more acutely stressful occurrences, like having the N-word shouted at me during a night out in my hometown in the Netherlands or being told I didn't "look British" by German airport police after losing my passport as a teenager. I had long understood individual incidents like these—sometimes euphemistically referred to as microaggressions—to be inconvenient and unpleasant but, until my conversation with Utsey, I hadn't really reckoned with the fact that they may collectively be harming my health.

"Racism is beyond an event," Utsey explained to me. "It's not just an event that happens to you. It's a physiological process that taxes your autonomic [nervous] system and causes physical illness. And this is really at the crux of health disparities," he argued. "Everyone's talking about, why are Black folks less healthy in a number of realms? Is it access to health care? Probably. Is it poverty that exposes them to more stress, poor diet, poor nutrition? Yeah. But I think the chronicity of race-related stress that creates this prolonged activation is also problematic," he said.

Utsey first began examining this issue from an academic perspective as a graduate student in New York in the 1990s. He was working as a counselor in Harlem at the time and noticed that the items on the Holmes-Rahe Life Stress Inventory—then a widely used scale among counselors to assess life stress—didn't reflect the lived experiences of his clients, many of whom were Black. "There was no reference to racism or police brutality," he told me. So, Utsey created his own scale to measure life stress among Black populations. He included examples of everyday racism that might induce stress like "You have been followed by security (or employees) while shopping in some stores" and "You called the police for assistance and when they arrived they treated you like a criminal." Since he published his scale, called the Index of Race-Related Stress or IRRS, in 1996, multiple studies by Utsey and others have linked chronic exposure to racism-related stress with reduced mental and physical well-being among Black people and people belonging to other marginalized racial and ethnic groups in the US and elsewhere.[133]

Over a lifetime, racism-related stresses and challenges may take a serious toll on both body and mind. In 2019, a team that included health disparities researcher Michele Evans at the National Institutes of Health found potential signs of this in the brains of a group of older African American people. Evans and her colleagues were investigating why Alzheimer's-related dementia is more prevalent among Black people compared to White people in the US and whether this might be linked to racism. Over a five-year period, she and her team surveyed a group of seventy-one African American study participants about their experiences of racism. They also used magnetic resonance imaging (MRI) to scan the participants' brains and measure something called white matter lesion volume—an early indicator of cognitive decline. They found that among older African Americans, increases in perceived lifetime discrimination burden were associated with increases in white matter lesion volume over the study period. Separate research has linked perceived frequent experiences of racism among African American women with self-observed declines in cognition.[134]

In other words, race may be a social rather than a biological construct, but racism clearly affects our biology.

In addition to the potential harm caused by the chronic humdrum of day-to-day discrimination, more acute and traumatic experiences of racism may also cause lasting damage to the brain. Emerging evidence points toward a link between past experiences of trauma and dementia risk in later life.

Research suggests that US veterans with post-traumatic stress disorder (PTSD) experience higher rates of dementia than those without PTSD, for instance. A study of first responders involved in search, rescue, and recovery during the 9/11 attacks on the World Trade Center in New York also found that those with PTSD experienced increased cognitive impairment and possible dementia compared to those without the condition. As of 2023, the American Psychiatric Association defines PTSD as "a psychiatric disorder that can occur in people who have experienced or witnessed a traumatic event, series of events or set of circumstances [such as] natural disasters, serious accidents, terrorist acts, war/combat, rape/sexual assault, historical trauma, intimate partner violence and bullying," and, although its website notes that people belonging to racial and ethnic minority groups—particularly "Latinos, African Americans, and Native Americans/Alaska Natives"—are "disproportionately affected and have higher rates of PTSD than non-Latino Whites," the definition of and criteria for PTSD have been criticized by some psychiatrists and psychologists in the past for their lack of attention to stressful events connected to racial discrimination and experiences of racism.[135]

Race-based traumatic stress—defined by psychiatrist Robert Carter in a seminal 2007 paper as "emotional trauma brought on by the stress of racism"—has been found to share some symptoms with PTSD, while racism-related stress more broadly has been associated with more severe PTSD outcomes. A more recent analysis that examined the existing scientific literature on racial discrimination and trauma in the US confirmed that there was a positive association between the two and called for more research in this area.[136]

But we don't need to look to scientific studies to understand the deep-seated and long-standing relationship between racism and trauma. A year after witnessing and capturing on camera the murder of George Floyd, by White police officer Derek Chauvin, in Minneapolis in May 2020, when she was just seventeen years old, Darnella Frazier wrote a social media post about the trauma the experience had caused her. "I am 18 now and I still hold the weight and trauma of what I witnessed a year ago," she wrote. "It's a little easier now, but I'm not who I used to be. A part of my childhood was taken from me," said Darnella. "My 9-year-old cousin who witnessed the same thing I did got a part of her childhood taken from her," she added.[137]

What started out as a normal day for Darnella, who had been walking her young cousin to the corner store, quickly turned into a nightmare when she came across George Floyd being arrested. She told a court in 2021 that she had started recording the incident on her phone, because she "saw a man terrified, begging for his life." We all know what happened next.[138]

I remember watching Darnella's video on my smartphone from my living-room-turned-office in Berlin and feeling sick. I felt even worse when I learned that the video had been filmed by a seventeen-year-old. Even watching a video like that can be extremely traumatic, let alone witnessing what happened firsthand and capturing it for the world to see. A survey of 134 mainly Black or Latinx college students in the US published less than two months after George Floyd was murdered found that the students, aged between 18 and 24, reported experiencing symptoms consistent with PTSD after viewing videos on social media showing Black men being killed at the hands of police officers.[139]

"Everyone talks about the girl who recorded George Floyd's death, but to actually be her is a different story," Darnella wrote in her social media post. "Although this wasn't the first time, I've seen a black man get killed at the hands of the police, this is the first time I witnessed it happen in front of me. Right in front of my eyes, a few feet away," she said. "It changed me. It changed how I viewed life. It made me realize how dangerous it is to be Black in America."

Several studies have suggested that experiencing trauma during childhood is associated with dementia risk later in life. In 2017, a team of researchers led by Kylie Radford at the University of New South Wales in Australia identified an association between childhood trauma and late-life dementia risk among Aboriginal and/or Torres Strait Islander peoples. Although the word "racism" isn't mentioned in their paper, Radford and colleagues mention that the study participants' scores in a childhood trauma questionnaire were associated with several indicators, including separation from family "by a mission, the government or welfare." Many First Nations children were forcibly removed from their families between 1910 and the 1970s, as part of a series of assimilation policies implemented by the Australian government based on racist and pseudoscientific notions of Black inferiority and White superiority. They proposed that First Nations children should be either allowed to "die out" or forced to assimilate into White communities. Mixed children, of First Nations and White parentage, were especially vulnerable to being separated from their families as it was thought that their perceived lighter skin color would make it easier for them to blend into White communities. The generations of children removed under these policies became known as the Stolen Generations, and the resulting legacy of trauma and loss continues to affect First Nations communities to this day—including participants in Radford's study—the results of which may at least partly explain why Aboriginal and/or Torres Strait Islander peoples experience a rate of dementia three to five times greater than Australia's wider population.[140]

Sleep disorders such as insomnia are also common following traumatic experiences. In her social media post on the one-year anniversary of George Floyd's death, Darnella explained how the trauma from witnessing his murder had impacted her sleep. She described "closing my eyes at night only to see a man who is brown like me, lifeless on the ground," recalling that she "couldn't sleep properly for weeks" afterward. "I used to shake so bad at night my mom had to rock me to sleep," she said.[141]

Sleep is a biological necessity. "If you're not getting a healthy amount, all sorts of things go wrong," sleep researcher Girardin Jean-Louis at the University of Miami in Florida told me. Failing to get between seven and eight hours of sleep a night has been associated with elevated risks of obesity, high blood pressure, diabetes, cardiovascular disease, and cancer, explained Jean-Louis. Yet sleep is a privilege. According to Jean-Louis, in the US, "Blacks as a whole group are sleeping 30 minutes less, 40 minutes less, compared with Whites." Indeed, a 2017 study that analyzed the sleep habits of 426 people from across the country found that Black people slept 40 minutes less than White people on average. It additionally found that Black people experienced poorer sleep quality than White people, spending 10 percent less time asleep while in bed, and that the lower amount of sleep and quality of sleep among Black people was associated with increased risks of cardiovascular disease and diabetes. Research by Jean-Louis and his colleagues, both in the US and the Netherlands, has also identified racial and ethnic inequalities in sleep health.[142]

Jean-Louis said that he spends a lot of time visiting churches, barbershops, and beauty salons—places at the heart of many Black communities throughout the US—to share his knowledge on the importance of sleep for health. His conversations with the people he meets there hint at some of the factors that may underlie the racial sleep gap.

"At the barbershop you will hear some people say, 'Well, gee, you know, each time I hear a siren behind me as a driver, I tense up, my blood pressure goes up.' Well, after a while, through conditioning, even if you're not driving—you may be at home—if you hear a police car [siren] going or if you see flashing lights going by, through conditioning, you may come to associate those things with something that causes high blood pressure," he said, noting that Black people are more likely than White people to be stopped by police while driving in the US. Such anxiety is unlikely to be helpful if you are trying to fall asleep.

Jean-Louis suspects that the higher levels of noise, light, and air pollution known to be present in areas where Black people and other people of color disproportionately reside may also be a contributing factor to sleep disparities in the US. Further, Black and Latino people are additionally overrepresented among workers who do night shifts, which is associated with poorer sleep health due to disruption of the body's internal clock or circadian rhythm and has been classified as "probably carcinogenic" by the WHO.[143]

Jean-Louis additionally touched on something else that, as a descendant of transatlantic slavery, I found particularly intriguing. "How does trauma associated with slavery contribute to poor sleep?" he mused. "We don't know. This is sort of the question we are trying to answer in our own lab," he said. "If slavery is in fact a traumatic event, much like any other traumatic event, it has long-lasting effects on folks, even generations after," he said. His work exploring possible health effects of trauma associated with the history of transatlantic slavery was at a very early stage when I spoke to him, but the question of whether and how the effects of trauma might be transmitted intergenerationally from parents to their children—and even to subsequent generations—has long stoked both scientific and general interest. And it has been explored scientifically in the context of another historical, racism-fueled atrocity.

Several studies have found evidence of a greater prevalence of PTSD among the adult children of Holocaust survivors in comparison with other adults, particularly among those whose parents also experienced Holocaust-related PTSD. Initially, this pattern was mainly attributed to the cultural transmission of trauma, for example through parental behavior or storytelling. But as our understanding of genetic inheritance improved around the turn of the century, another possible explanation emerged. Namely, that trauma could cause changes to our biology, which might then be transmitted to our offspring. "More and more, we are seeing that trauma can in fact change the genetic makeup," Jean-Louis told me. "Epigenetics plays a role in some of our health, and we're learning more and more about this," he explained.[144]

Epigenetics is the study of changes to DNA that don't affect the underlying DNA sequence but which can influence how our genes work, for instance, by determining whether or to what extent a particular gene is switched on or off. These epigenetic changes are chemical tags or marks, which are added to or removed from the DNA in response to environmental triggers. In the 2000s and 2010s there was a huge amount of scientific and general excitement about the possibility of transgenerational epigenetic inheritance—the idea that epigenetic changes could be passed on to subsequent generations. In other words, that environmentally induced changes to our DNA might be heritable. Several studies demonstrated the phenomenon in animals, such as mice, while others provided indicators that it might happen in humans too.[145]

During the same time period, it was becoming clear that specific patterns of epigenetic changes are associated with certain health conditions, including psychiatric conditions such as PTSD. As this emerging scientific field exploded, some researchers began to wonder whether epigenetic inheritance might therefore help to explain observations of apparent intergenerational transmission of the effects of trauma. Rachel Yehuda, a psychiatry and neuroscience researcher at Mount Sinai School of Medicine in New York, was one of them. She set up a clinic for Holocaust survivors in the 1990s and was among those who first noticed and published findings on the pattern of increased rates of PTSD in their children.[146]

Fast-forward a couple of decades to the epigenetics era, and Yehuda and her colleagues had made another intriguing discovery. In a small study they found, to great general interest, that epigenetic changes associated with trauma were present on the same stretch of DNA in both

Holocaust survivors and their children. The stretch of DNA in question sits within a gene called *FKBP5*, which is involved in the body's response to stress. In their paper, Yehuda and her team concluded that their findings provided "a potential insight into how severe psychological trauma can have intergenerational effects" and speculated that these effects might have been mediated by changes to the parental egg or sperm cells that formed the offspring. But they acknowledged that this couldn't be definitively determined based on their results, since they didn't directly analyze DNA from sperm or egg cells in their study, and they highlighted that it would be necessary for future studies to look at multiple generations in order to differentiate epigenetic versus socially mediated effects.[147]

In fact, teasing apart the effects of sociocultural factors versus any potential epigenetic mechanisms by which trauma might be transmitted down generations is increasingly proving to be, at best, extremely scientifically challenging and, at worst, impossible. For example, in addition to being affected by their parents' trauma through altered parenting behavior and storytelling, the children of Holocaust survivors have undoubtedly also been affected simply by virtue of belonging to a marginalized group. After all, we have just seen several examples of how chronic exposure to racism—including antisemitism—can harm health. This becomes even more complex with subsequent generations, such as the grandchildren and great-grandchildren of Holocaust survivors or even the many generations of descendants of transatlantic slavery, whom Jean-Louis referred to earlier.

Academic and author Joy DeGruy used the term "Post Traumatic Slave Syndrome" in her 2005 book by the same name to encompass the complex and multiple ways in which slavery and continued discrimination against Black people manifest as intergenerational psychological trauma among Black Americans. "Ten years ago, people were like, 'Oh my God, that's ridiculous,'" said Dr. Alfiee, referring to DeGruy's theory. "It's only been recently that more people have come around to understand that what she's talking about is, you had, what, four hundred years of Black people having no rights, being treated as subhuman, right? Denigrated, raped, murdered, all kinds of things," Dr. Alfiee continued. "In that period of time, no one has ever provided any kind of universal mental health care for those of us who are descended from people who were enslaved," she pointed out. "I think it has a direct impact [on health]," she said.[148]

That impact may not be mediated through epigenetic inheritance, but I don't think that makes it any less real. I think that people experiencing trauma related to racism, including traumatic racist events in the past or present, deserve to feel validated and supported.

The effects of interpersonal racism on health are perhaps more challenging to record, measure, and quantify, compared to those of systemic racism, but that doesn't mean they aren't worth examining. These two broad categories of racism are complex and overlapping—as are their impacts on health. In fact, I would argue that it is almost impossible to understand the full impact of interpersonal racism without considering it in the wider context of systemic racism and societal inequality. This comes across in Clint's telling of the racist incident he witnessed at the airport in Charlotte, North Carolina.

Clint had locked eyes with the Black woman as the White woman vanished into the crowd in the terminal, her racist language still hanging in the air. "I think we were both processing what had just occurred, how quick this woman had been to wield that word as the weapon she knew it was, and how quickly she had then run away," he recalled in his 2022 *Atlantic* article. Afterward, he and the Black woman who had been the target of the racist abuse reported what had happened to a staff member, but by then the White woman had already disappeared. With the benefit of hindsight, Clint wondered whether he should have done anything differently. "Should I have responded faster? Should I have confronted the woman? Should I have stood in front of her to block her way until an airport official came over? But what would I have been hoping to achieve in that? For her to miss her flight? For her to be placed on a list? For her to apologize?" he asked. "Then I imagine the optics of a Black man attempting to physically prevent a smaller white woman from leaving, and immediately recognize the way that such a move would create its own spectacle, its own dangers," he added.

What this demonstrates is that Clint and the Black woman standing beside him didn't experience the racist incident at the airport in isolation, they experienced it within the wider context of a society in which there are significant power dynamics based on race. Those power dynamics, when combined with the daily microaggressions and acute instances of racism or racial trauma we have just explored, produce a perfect recipe for poor health among people belonging to marginalized racial and ethnic groups.

Of course, systemic and interpersonal forms of racism within societies overlap to harm health. A clear example of this is colorism—a form of racism where people within a particular racial or ethnic group are discriminated against based on skin tone or other racialized features, such as hair texture. And that's where my investigation took me next.

CHAPTER 5

Colorism

For most of my life I didn't feel I could wear my hair the way it naturally grows from my head. As a child, I already stood out from the crowd at school because of my brown skin, but the thing that I most remember being bullied about was my hair. My mother tried to protect me: from the age of about five, she started taking me to the hairdresser to get my hair relaxed, meaning chemically straightened. Later, she began buying the products herself and relaxing mine and my younger sister's hair in our bathroom at home. "I wanted my kids to be accepted, to fit in," she later explained to me.

Growing up in Lebanon in the 1960s and '70s as the only Black child in her family, and one of only two Black children at her school, my mother faced significant discrimination. She is Lebanese on her father's side, and Ghanaian on her mother's side. Her parents divorced when she was a baby, in Ghana, and she eventually ended up in the care of her father's family in Lebanon. He soon got remarried to a Lebanese woman—my *teita* (grandma in Arabic)—and together they had five more children. Being the only Black or Mixed child in her lighter-skinned family was tough for my mum. During family outings, people found it difficult to believe she was part of the family.[149]

Her hair was also an issue for people. Her family didn't know how to take care of textured hair, so they always kept her head closely shaved. She was bullied about her hair at school, further lowering her already low self-esteem. This was all the more reason why she was delighted, as a teenager, when she learned of a product that could make her hair straight.

"I was desperate to fit in," she told me. "I had no idea what it was and what it involved," she said, remembering her first hair relaxing appointment. "It was an unpleasant experience." In those days, she recalled, the hair relaxing products available were very thick pastes that were difficult to spread onto the hair—and they didn't smell great. The process could also be painful. "I started experiencing this tingling and burning," my mum said.

Listening to her describe those sensations brought back memories of having my own hair relaxed. I would sit patiently with my scalp on fire because there was this notion that the longer you left the product on your hair, the straighter it would become. Eventually, I would be relieved by the flow of water against my hair and scalp, removing the foul-smelling product and leaving me with what I hoped would be a slightly more acceptable appearance. Looking back, I think a part of myself was also being washed down the drain, along with those nasty-smelling chemicals (and a bunch of money).

"I feel saddened and angry that we were made to feel inferior, because of frizzy hair," my mum said. Even though she experienced "a few little burns" on her scalp from relaxers on one or two occasions, she thinks the emotional scars from that feeling of inferiority affected her more.

She and I were both wearing our hair naturally as we sat together in her living room, sharing experiences. I stopped relaxing my hair when I was in my early twenties, in the early 2010s; the natural hair movement was taking off at the time, encouraging people like me to embrace our natural Afro-textured hair. My mom and sister soon followed. My mom was in her early fifties when she stopped relaxing her hair. "Seeing you and your sister embracing your natural hair really impacted me," she said, tears in her eyes. "It makes me very proud."

I don't judge my mom for choosing to relax my hair when I was a child, nor myself for continuing to do it as a teenager and young adult.

There are plenty of reasons why millions of people with textured or Afro hair globally use products to chemically straighten it. But I think a lot of those reasons are related to racism—or, more specifically, colorism.

The author and activist Alice Walker is often credited as being the first person to use the word colorism, which she defined in her 1983 book *In Search of Our Mothers' Gardens* as "prejudicial or preferential treatment of same-race people based solely on their color." Colorism is usually associated with preference for lighter skin tones and prejudice against darker ones. In practice, colorism often also encompasses featurism—a less frequently used term that refers to similar prejudice based on commonly racialized features, such as nose or eye shape and hair texture. At its root, colorism is a form of racism—but while racism separates people *into* racial and ethnic groups, colorism typically separates people *within* those groups. Like racism, colorism manifests both at a systemic and an interpersonal level, and causes harm to mental well-being as well as to physical health.[150]

To illustrate this, let's continue with the example of hair discrimination. My mom told me that when she worked as a pharmacist in London in the 1980s and '90s, even though she was part of a diverse team, she didn't feel that natural hairstyles would be considered acceptable in her workplace. "There was nobody who was going with natural hair," she said. A few decades later, when I was living in London myself, I was preparing to travel to Oxford for a PhD interview when a Black friend recommended that I should straighten my hair before leaving. "You can't show up with an Afro," she said. "It isn't professional." I am proud to say I showed up with my Afro anyway and I was accepted into the PhD program but, unfortunately, many people aren't so lucky.

In 2016, about a year after that PhD interview, Google's image search feature came under criticism after a user tweeted screenshots revealing that a search for "unprofessional hairstyles for work" retrieved mainly images of Black women with natural hairstyles. Typing "professional hairstyles for work" into the search bar, on the other hand, yielded mainly photographs of White women with straight hair. According to the Halo Collective, an advocacy group fighting against hair discrimination in the UK, one in five Black women feel societal pressure to straighten their hair for work, and 46 percent of parents say their children's school policy penalizes Afro hair. Ruby Williams, a pupil in London who was repeatedly sent home from her school in Hackney because of her Afro hair, received $10,390 in an out-of-court settlement in 2020 after her family took legal action against the school (the school didn't accept any liability). There have been similar instances of hair discrimination in other countries, including the US and South Africa. In the US, the last few decades have seen numerous court cases filed by Black employees alleging workplace hair discrimination and, as of 2022, at least eighteen US states have felt the need to pass legislation known as the CROWN Act to explicitly make hair discrimination illegal. Even Olympic athletes aren't exempt from this form of racism. In July 2021, the international swimming federation FINA turned down an application from the Black-owned swimming cap company Soul Cap to use their products at the Tokyo Olympics that year. Soul Cap is designed to protect Afro or textured hair, as well as natural hairstyles such as braids, twists, and locks.[151]

If discrimination at work and at school wasn't enough to persuade someone to try chemical straightening, worrying about their safety might. The first time I remember being called the N-word in a threatening manner, by a group of guys outside a nightclub in the Netherlands, they used my hair to target me. I hadn't touched up my roots for a while, so even though most of my hair was still relaxed, it looked more textured than usual. "N★★★★★ with the Afro!" I remember them shouting. I was a teenager at the time and I felt afraid. I had lost track of my friends, so I was on my own. I rushed home as fast as I could, my heart racing.

We have already seen how racism in societies contributes to disparities in general health and well-being for marginalized racial and ethnic groups. But the pressure for Black women, in particular, to solve the problem of hair discrimination by using products like relaxers comes with its own very specific set of potential health risks. Chemical burns are an example. "Some women do actually suffer third-degree burns, chemical burns," said Kimberly Bertrand, an epidemiologist at Boston University School of Medicine in Massachusetts whom I spoke with a few weeks before my conversation with my mother.

Bertrand and her colleagues published a study in 2021 examining data on hair relaxer use and breast cancer incidence among a group of more than fifty thousand Black women in the US. She emphasized that the main takeaway from that study was that moderate use of hair relaxers wasn't associated with an increased risk of breast cancer. "Our findings, I would say, were really quite reassuring," said Bertrand. "For most women in our study, there was no evidence of a link between use of these relaxers and future breast cancer," she explained. However, Bertrand and her colleagues did notice a slightly concerning pattern among a subset of the women in the study who reported using hair relaxers containing a chemical known as lye. They found that the women who reported using lye-based relaxers most frequently and over the longest periods of time—more than seven times a year for a period of more than fifteen years—were 30 percent more likely to develop breast cancer, compared with those who reported lighter use of these products. "That's a concern for sure," said Bertrand.[152]

The data Bertrand and her colleagues analyzed were from the Black Women's Health Study, which began tracking the health of fifty-nine thousand self-identified Black women in the US in 1995, with the hope of identifying factors contributing to racial disparities in health. "Most women were recruited from subscription lists to *Essence* magazine," said Bertrand, who first became involved in the study as a researcher back in 1997. The enrolled women, who were aged between twenty-one and sixty-nine at the time the study started, regularly complete surveys and share health data to provide a resource for researchers to identify potential associations between various environmental factors and health outcomes. Bertrand is particularly interested in breast cancer.

"Black women are 40 percent more likely to die from their breast cancer than White women," she noted, highlighting the glaring disparity we touched on in chapter 2. "In the US, you can imagine that there are lots of factors that contribute to that higher mortality. Much of it is, I would say, racism," she said. One aspect of that racism is hair discrimination, which, as we just explored, prompts many Black women to opt for chemical hair straightening.

"Relaxers are heavily marketed to Black women," said Bertrand. She explained that after coming across earlier research highlighting a potential link between exposure to some of these products and breast cancer risk, she had decided to investigate further. "Chemical hair relaxers or straighteners, as we call them, we know they contain many thousands of chemicals," she told me. "And we know they contain chemicals that are known to be toxic, especially these toxic chemicals called endocrine disruptors," she said.[153]

Endocrine disruptors are chemicals that can interfere with signaling by hormones, such as the sex hormone estrogen, which plays a key role in many breast cancers. These chemicals may have profound effects on our bodies. A 2011 study led by Tamarra James-Todd, an epidemiologist now at Harvard University, found that African American children were more likely to use hair products and to start menstruating at younger ages compared to children belonging to other racial and ethnic groups in the US. A more recent study by James-Todd and colleagues showed that several hair products commonly used by Black women exhibit hormonal activity in laboratory tests, suggesting that they could potentially contribute to reproductive and metabolic health disparities. And although Bertrand's research suggests that moderate use of hair relaxers in general doesn't increase breast cancer risk, she believes the observation of greater risk among heavy users of lye-based products in particular warrants further investigation. "I don't know if this is why these women got cancer. But it does make you think, what can we do as a society to make these products safer, or to counsel women about the risks in their own daily lives?" she said.[154]

When my mother and I were swapping stories about our experiences with hair relaxers a few weeks later, I thought back to my discussion with Bertrand. At the time when my mom started relaxing her hair, the range of products on the market was quite limited. She remembers all the products available to her being lye-based ones. By the time she started taking my sister and I to have our hair relaxed as children, the range of available products had expanded, and many were non-lye relaxers. In fact, my mom said that the products that were used on my and my sister's hair were specifically marketed as being for children. "When I was young, it was not promoted for children. There was no such thing as a relaxer for children, you had to be a certain age before you could use the relaxer. But when you were kids, suddenly there were these products," she told me. I asked her if she had ever been concerned about the potential health risks associated with any of the ingredients in these types of products, particularly the lye-based ones that she had used on her own hair for many years. "No," she answered. "It never occurred to me about the chemicals in the products, but I was concerned about the burn. I thought, this can't be good for your scalp to have repeated burns," she added, though I couldn't help but worry about the potential harm done beyond her scalp. I resolved to refresh my memory on what lye actually is—and to try to understand the basic chemistry of hair and hair relaxing.

Lye is a common name for the compound sodium hydroxide. I am familiar with sodium hydroxide in its pure form from my days as a laboratory researcher. It is very alkaline and corrosive, so in the laboratory we always worked with it in a chemical fume closet for safety reasons, in addition to wearing personal protective equipment such as gloves and safety goggles. Sodium hydroxide is also the main ingredient in many drain cleaners, which, incidentally, have pH ranges similar to those of many hair relaxers. In the case of the cleaning product, the sodium hydroxide helps to decompose hairs and solubilize fats clogging the drain by reacting with them to make soap. With hair relaxing, on the other hand, the goal isn't to completely decompose the hair but rather to break the strong chemical bonds that exist naturally between the keratin molecules in Afro or textured hair. All human hair, regardless of texture, is made up of the protein keratin. But in Afro or textured hair, strong chemical bonds called disulphide bonds connect the keratin molecules together in a way that produces kinks or curls. By breaking these bonds, hair relaxers leave the hair permanently straight. But just as hair dye can't permanently alter hair color as new hair grows, hair relaxers can't permanently change the texture of the hair that grows out from the scalp. In the past, that's what kept my mom—and later me and my sister—going back to the salon every so often to touch up our roots.

In lye-based relaxers, the ingredient responsible for the straightening effect is sodium hydroxide, whereas in non-lye relaxers, other compounds—such as calcium hydroxide, which is slightly less alkaline than sodium hydroxide—are used instead. These are the chemicals that are responsible for that burning or tingling sensation that my mother and I talked about feeling on our scalps, during hair relaxing. They are also the reason why relaxers can cause damage to the scalp, including cuts and burns, as my mom experienced. But relaxers also contain plenty of other chemicals, often including endocrine disruptors—the chemicals Bertrand cautioned about—which aren't always specifically named in ingredients lists because of the proprietary nature of formulations. For instance, "fragrance," which is listed as an ingredient in one of the hair relaxer products I used to use, can refer to many different chemicals, including endocrine disruptors.[155]

Bertrand suspects that lye, which is particularly harsh, may damage the scalp in such a way that it makes it easier for other chemicals in the hair relaxer to then penetrate the skin and potentially cause further harm. "Our skin is a big component of our immune system. The purpose of our skin is to serve as a physical barrier to the outside world," she said. "If you're getting scalp burn on your head, and creating wounds, that's a mechanism of entry for the hair straightener to enter your bloodstream, and these other chemicals, the known endocrine disruptors or potentially other toxic chemicals, to get into your system," Bertrand explained.

Even in the absence of a visible burn or wound, the lye could still be causing microtears to the skin, she pointed out. An increased penetration of endocrine-disrupting chemicals through the scalp and into the bloodstream following use of lye-based relaxers could potentially explain why the heavy users of these products in Bertrand's study had an increased risk of breast cancer.

Bertrand thinks the contents of hair relaxers should be more tightly regulated, for instance by the FDA. "I consider this an issue of environmental justice," she told me. Indeed, the vast majority of people who use hair relaxers around the world are Black. My mother agrees—she suspects that the fact that the main users of hair relaxers are Black women may have something to do with the limited amount of research and regulation in this area, particularly at the time when these products first became available on the market. "People didn't care to do the research," she told me. In the European Union, where my mom and I both currently live, manufacturers are legally obliged to ensure that cosmetic products undergo an expert scientific safety assessment before they are put on the market, but in the US there isn't a requirement for ingredients in cosmetics to be safety tested. Even in the EU, there is pressure for legislators to do more. In 2021, the campaign group EDC-Free Europe called for the EU to completely ban the use of endocrine disrupting chemicals in cosmetics, for example.[156]

For my mother, all of this is too little too late. Sitting together in the living room, we both wondered what we might have done differently if we had been more aware of the potential health risks associated with hair relaxers. Would my mom still have opted to chemically straighten her hair all those years? We agreed that she might have, because the pressure to feel accepted by society was that strong. "I think if I was accepted and liked for who I am, it would never have occurred to me to make any changes about how I look," she said. "I think the major factor in me having to do things like that is the fact that I was discriminated against and there was racism, and I wasn't accepted as I am," said my mom.

When it comes to racial disparities in breast cancer mortality in the US, as Bertrand put it: "Society is not off the hook." While hair relaxer use may contribute minimally to that racial health gap, its contribution is most certainly dwarfed by other factors driven by structural racism, such as economic and environmental inequalities, she pointed out. Bertrand thinks the lesson from her research on hair relaxers might simply be "everything in moderation" and she hopes that increased research on the ingredients in these products might help to pinpoint any particularly harmful substances, which could then be removed. "It's not my advice to stop using them," she told me. "I think women should feel empowered to wear their hair how they want to wear their hair."

Colorism and featurism operate within, as well as between, racial and ethnic groups, with wide-ranging health harms. For instance, even within the natural hair movement, supposedly aimed at encouraging Black people to love their hair in its natural state, some hair types are more celebrated than others. When I stopped chemically straightening my hair, I turned to social media channels for guidance from natural hair beauty bloggers on how to care for and style my natural hair. But I quickly noticed that hair textures with looser and more defined curls—known in the natural hair community as "type 3" hair—seemed to be perceived as more favorable than kinkier "type 4" hair, like mine. This reminded me of when I used to go to get my hair done with my family as a child and teenager: the Black hairdressers would often comment on the fact that my hair was "bad" or "difficult" in comparison to my mom's or my sister's, because it was kinkier than theirs. Growing up, people also compared our skin tones, with the implication being that lighter was better. This skin color hierarchy was promoted by White people too, and not always in the most subtle of ways. A former employer of mine in the Netherlands once commented while I was working that he thought my skin color was beautiful, but that anything darker wasn't appealing to him (it was intended as a compliment). As soon as he made that comment, I thought back to one of my first visits to Ghana with my family when I was a child. Looking out of the car window as we drove around Accra, I noticed that most of the people on billboards and in other advertisements had skin tones several shades lighter than the average person on the street. Later, as an adult, I would realize that some of those ads were for products promoted as skin lighteners (and that even those ads for other products were reflecting the same beauty standard).

Unlike hair straightening, skin lightening wasn't a thing in my family when I was growing up. "I wasn't in a situation where I considered that," said my mother, when I asked her about her experiences as a young person. When my mom was fourteen, civil war in Lebanon forced her and her family to flee to Ghana, which is where she first came across skin lightening or bleaching cream. "I remember seeing people in Ghana who were using it, and I knew it burns and damages your skin," she said. Although her childhood in Lebanon had taught her that her brown skin was undesirable, she decided it wasn't worth it to risk burning her face in the pursuit of lighter skin. "If somebody told me, oh, there's a pill, if you swallow it, that makes you White? Who knows what I would have done then?" she commented.

I could relate. As a child in the Netherlands, I remember wishing that I was White and blond. However, it never occurred to me to try to lighten my skin. Perhaps if I had been living in Ghana, where skin lightening products are more heavily marketed, I might have considered it. In the end, I think my and my mother's decisions not to lighten our skin are probably also reflections of our relative privilege. We both naturally have lighter brown skin tones, not too far off those promoted as the desirable end-result in advertisements for skin lightening products in Ghana. If we had darker brown skin, we both might have felt more pressure to lighten. Indeed, many people in Ghana and in other countries around the world *do*. This isn't about vanity; in many cases people feel that having lighter skin will increase their economic and social opportunities. There is evidence to support this. Research from the US has shown that Black job applicants with fairer skin are considered more favorably than equivalent Black applicants with darker skin tones, for instance. Meanwhile, in India, you only have to look at the casting in Bollywood to notice the strong bias toward fairer-skinned people. This bias extends to marriage opportunities too. A study on arranged marriages in India found that darker-skinned marriage candidates were rated as less preferable by their prospective in-laws, compared to lighter-skinned candidates. Colorism is present in countries all over the world, and given the impact that it has on people's life opportunities, it is wholly unsurprising that the global market for skin lightening products was estimated to be worth $8 billion as of 2020. All of this comes at an equally global cost to health.[157]

"Here in the Philippines, skin whitening is really a big business," Thony Dizon, a chemical safety campaigner at the Ecological Waste Coalition of the Philippines (EcoWaste Coalition) advocacy group, told me. Dizon is extremely passionate about his campaign work raising awareness of the health risks associated with skin lightening—or whitening, as he calls it—in the Philippines. In particular, he has long been concerned about illegal skin lightening products being sold at markets in the country and online, which often contain levels of mercury that are harmful to health. These products are popular because they are relatively cheap and can produce the effect desired by so many; mercury salts inhibit the production of melanin, the collective term for a large group of molecules responsible for skin pigmentation, thus lightening the skin.

The fact that mercury is toxic hasn't curbed the growing appetite for these products. "The toxic trade of often illegal mercury-added skin lightening products is a global crisis expected to only worsen with skyrocketing demand, especially in Africa, Asia and the Middle East," reads a 2019 report by the WHO. That report goes on to list the many adverse health effects of mercury in skin lightening creams and soaps, including kidney damage, skin rashes, skin discoloration and scarring, reduction in the skin's resistance to bacterial and fungal infections, anxiety, depression, psychosis, and peripheral neuropathy. Yet, as Dizon explained to me, even in countries such as the Philippines where these types of products are banned, they are still advertised and available to consumers via the internet as well as at street markets.[158]

"Our periodic visits to the marketplace have confirmed that the illicit trade in skin whitening cosmetics laced with mercury is quite widespread," said Dizon. He and his team regularly take samples from skin lightening products sold at markets throughout the Philippines or online and screen them in their laboratory for the presence of heavy metals, such as mercury. "We'll visit places where ordinary people will go to buy these personal care and cosmetic products. We'll buy samples and, once we purchase it, we screen it," he explained.

Whenever Dizon and his team identify any unsafe products, particularly those containing mercury above the trace amount limit of 1 part per million established by a global convention in 2017, they immediately notify the Philippines Food and Drug Administration (FDA). At least eighty such products were banned by the FDA between 2010 and 2018, as a direct result of his and his team's efforts. Dizon and his collaborators also work to share data regarding the health risks of these types of products to the wider public, since the products usually aren't labeled as containing mercury. "We create media stories using this data generated to inform the public," he said. The data are shocking. One analysis by the EcoWaste Coalition found that of sixty-five samples purchased from online sources in 2020, forty had mercury levels above the trace amount limit of 1 part per million, making them illegal to sell in the Philippines. The product with the highest level of mercury was found to have a concentration of 41,200 parts per million—more than 40,000 times the limit. "Imagine," sighed Dizon, as he shared the appalling figures with me. "The concentration of mercury is really very high."

Through his campaign work in the Philippines, Dizon has met many people who have experienced negative health effects as a result of exposure to mercury in skin lightening products. Some of these people have even joined his campaign, sharing their testimonies to try to help raise awareness about the dangers of these types of products. Grace Reguyal is one of them. Dizon connected the two of us, and Grace kindly agreed to speak with me about her experience. Although she understands English, she felt less confident with speaking, so her husband, Ben, joined us on the call to help interpret. Their young son appeared in the background of the video call as we were chatting. "Our son was also affected by this skin disease that was caused by this product," Ben told me. The product he was referring to is a skin lightening product, consisting of a soap and a cream, which Grace had decided to try out back in 2018. And the "skin disease" was in fact a reaction to mercury, later found to be present at illegal levels in the skin lightening product that Grace had used.

A friend of hers had used the product first and recommended it. Grace was drawn in by her friend's claim that it could whiten skin within just seven days. She ordered it online to try it out for herself and was delighted when her friend's claim proved true—her skin became visibly lighter. She had been using the soap and cream for more than three weeks when she first noticed some red spots had appeared on her skin. She wasn't especially concerned initially, putting it down to a mild heat rash. But within two days, the rash on her face, neck, and chest had become redder and more pronounced. It was also becoming itchy. Eventually, Grace and Ben noticed that their son, who was being breast-fed by Grace at the time, was starting to develop a similar rash. They had already begun to suspect that it might be related to the skin lightening product, and when they noticed that their son had also been affected, they became even more concerned. They took their son to see his pediatrician, and Grace also went to see a dermatologist, bringing samples of the skin lightening product along with her.

The dermatologist immediately confirmed her and her husband's suspicions, noting that Grace was the fifth patient who had come in recently with similar symptoms after having used the exact same product. The product was found to contain levels of mercury above the trace amount limit and should never have been sold to them in the first place, because it was illegal, the doctor explained. Fortunately, Grace and her son both recovered, although Grace still occasionally experiences skin rashes, especially when it's very hot outside or when she's stressed. The experience also caused her a great deal of mental anguish. When I asked her how it made her feel, she replied to me directly in English. "Depressed," she said.

Grace decided to share her experience publicly, via social media, in the hope that it might prevent the same experience from happening to someone else. She was invited to appear on several television programs in the Philippines, along with other people who had experienced similar negative health effects after using skin lightening products found to contain mercury. A photograph showing the rash on her face, neck, and chest, which she posted on Facebook, is featured in a report by the EcoWaste Coalition, the advocacy group that Dizon campaigns with. Dizon hopes that the publicity surrounding the experiences of Grace and others will help to raise awareness about the health risks associated with mercury-tainted cosmetics. In their report, the EcoWaste Coalition made several recommendations, including that the government should strengthen laws and regulations on the sale of these products, that the cosmetics industry should move away from White-centric beauty standards and instead promote diversity in beauty, and that online shopping and social media platforms should crack down on the sale of illegal products via their platforms. They also recommended that consumers should embrace their natural skin color with pride and stand up to color-based bias, prejudice, and discrimination.[159]

Dizon believes that colorism is a major reason why skin lightening—with all its associated health risks—is so prevalent in the Philippines. When asked in an internal survey in 2020 to suggest possible reasons for the popularity of skin whitening nationally, the following statements were among the top answers provided by EcoWaste Coalition members based across the country: "White skin is beautiful," "White skin is attractive," "Because it can be seen on TV, movies and commercials," "This is due to colonial mentality," "White skin is clean to look at," and "White skin symbolizes a better status in life." Dizon is passionate about tackling colorism in the Philippines to reduce the pressure on people to expose themselves to potentially dangerous chemicals in skin lightening products. "I like your color, Layal," he said to me at one point during our conversation, smiling. "Thank you. I like yours too," I replied, and we both laughed.

"We want Filipinos to embrace our natural skin color and stop poisoning our bodies," Dizon continued, speaking emphatically. "And imagine, Layal, it doesn't just poison our bodies, it poisons the environment," he pointed out. Mercury in skin lightening products, such as soaps and creams, is eventually discharged into wastewater. In the environment, the mercury becomes methylated and can enter the food chain, for instance via fish as highly toxic methylmercury (exposure to methylmercury during pregnancy can affect developing fetuses, resulting in neurodevelopmental conditions).

Back on the call with Grace and her family, I wondered whether their feelings about skin lightening products had been influenced by their ordeal. Grace still uses products to try to lighten her skin, but she is now much more cautious about the specific products she uses and where she purchases them. She said her motivation for pursuing a lighter skin tone is to boost her confidence, because lighter skin is seen as more beautiful in the Philippines. According to her husband, Ben, lighter skin is also often perceived by employers in the country as more presentable for the workplace. I asked him if he also felt pressure to lighten his skin. "Me, personally, I don't really mind the color of my skin. If I get dark, it's okay, because I'm a man," he replied.

There certainly seems to be a gendered aspect to colorism and skin lightening, as there is with the use of hair relaxers and with many other beauty ideals, although the practice is by no means restricted to women, according to Dizon. "Both men and women are using skin lightening," he emphasized. Data on skin lightening product use suggest that the balance is tilted more toward women, though. A survey of university students in twenty-six countries across Asia, Africa, and the Americas conducted in 2013 found that 30 percent of female and about 17 percent of male students reported having used skin lightening products in the prior twelve months, for instance.[160]

For Grace, marketing through online advertisements, as well as seeing many lighter-skinned TV personalities and social media influencers, contributed to her aspiration for lighter skin. She wanted to offer some advice to young women in a similar situation to her. "To everyone who's trying to look more beautiful, Whiter, just make sure that what you're buying is something that is approved, safe, effective," she advised. Nowadays, she only buys products from pharmacies, and she opts for well-known brands. The advantage is that these products should be safer than many of the cheaper, potentially mercury-laced products available at street markets or online, but the disadvantage is that they are also much more costly. "It's really more expensive," commented Grace's husband, Ben. Indeed while Dizon welcomes increased consumer consciousness regarding product safety, he worries about the many people in the Philippines who simply can't afford to buy pricier products. With the pressure to be Whiter so great, many people may feel left with no choice but to try cheaper and riskier options to lighten their skin.

It was becoming clear to me that the health and environmental injustices associated with skin lightening are direct products of racism, and specifically colorism. But when I put my interpretation to Grace and Ben, it didn't seem to resonate with them. "Here in the Philippines, colorism's really not that . . . that's not much of an issue," said Ben. So, if colorism isn't the culprit, why the perception of lighter skin as better or more beautiful, I wondered. According to Grace and Ben, "it's just a feeling." I understand that feeling. It's the same feeling I had as a child and teenager when I sat at the hairdresser with my scalp on fire, hoping to achieve flowing, straight hair. If I had been asked at the time to articulate why I thought straighter hair was superior to my natural Afro hair texture, I would have struggled to explain. In fact, it is only through the process of writing this book that I have been able to find the vocabulary to fully express what I was experiencing. It was the effect of racism, and specifically colorism, which I had internalized in the form of Eurocentric beauty standards. Whether or not we are conscious of it, this kind of internalized racism—and the extent to which it can drive us to try to alter our hair texture, skin color, and other racialized features—can cause harm that extends far more than skin deep.

"I think that it's so normalized that there's almost this lack of awareness and when you bring up this issue of colorism, people are almost offended," said Joanne Rondilla, a sociologist at San Jose State University in California. Rondilla's research explores the complex ways in which colonial legacies exist within our everyday lives, including in the context of power dynamics related to skin tone. As part of her research, she has conducted extensive interviews of people both in the US, where she's based, as well as in the Philippines, examining the relationships between historical colonization, colorism, and skin lightening. Her research interests have very much been shaped by her own experiences of colorism as a Filipina born and raised on the US island territory Guam, and as someone who spent over twelve years working in the cosmetics industry, which has long been criticized for promoting Eurocentric beauty ideals.

"These ideas of beauty definitely damaged me," Rondilla told me. "So much so that I ended up researching it," she said with a chuckle. Growing up, Rondilla was often compared to her mother who had lighter skin than her. She quickly learned from her family and wider community that lighter skin was valued over darker skin—an idea that was reinforced during her time working in the cosmetics business. "I interviewed someone, and I'll never forget, you know, this thing that she said. She said, 'In the Philippines, you are not a cosmetics company unless you have a skin lightening product.'" Rondilla feels companies that profit from colorism, by marketing and selling skin lightening products, are perpetuating the problem. Even though the skin lightening products marketed by large cosmetics companies might not be unsafe in and of themselves, the promotion of skin lightening in general undoubtedly feeds into the narrative that lighter skin is more beautiful and desirable, contributing to demand for more dangerous, illegal products. And selling lighter skin as a beauty ideal isn't only harmful to physical health, Rondilla told me, it harms mental health too.

"Here's an example," said Rondilla. "One of my interviewees had talked about how, because she was the darkest daughter, as well as the darkest cousin, [. . .] her mom often harassed her about her skin tone." The same interviewee told Rondilla that in addition to encouraging her to use skin whitening soaps, when she was eight years old her mother had once rubbed lemon juice all over her skin in the hope that it would make it lighter. "She decided to do that to me not knowing how bad it would hurt. "'Cause the lemon juice, you know, is very acidic," said the interviewee. "She felt she was being punished for her skin," said Rondilla. The pain was emotional as well as physical. In the words of the interviewee: "It's traumatizing when people tell you to be a certain way. Even though you really can't control who you are."[161]

Increasing general awareness about the history behind colorism—and its harmful health effects—may help to tackle the problem. Before speaking with Rondilla, I was aware that colorism within Black communities globally had been traced back to skin color hierarchies established by European colonizers and enslavers in Africa and the Americas. Enslaved people received differential treatment according to skin tone: while darker-skinned enslaved people in the Americas were usually forced to labor outdoors in fields, those with lighter skin were typically made to perform domestic tasks indoors. This differential treatment has been linked to the fact that lighter-skinned enslaved people were often related to their enslavers; rape of enslaved women by enslavers was widespread. European colonization in Africa and Asia also contributed to colorism in the affected countries. In India, for example, British colonial rule is thought to have exacerbated colorism that has been connected to the country's preexisting, ancient Hindu caste system. Marginalized caste groups include people sometimes referred to as Dalits or "Untouchables," who were excluded from the traditional Hindu caste hierarchy—an occupational division of people, where ancestry determines acceptable occupations and positions in society—and considered to be impure due to their perceived darker skin and lower-status occupations.[162]

Despite modern laws prohibiting caste discrimination, persistent structural inequities and prejudices continue to contribute to poorer health outcomes among marginalized caste groups, according to Suresh Jungari, who researches social determinants of health at the International Institute for Population Sciences in Mumbai, India. Jungari, who grew up more than six hundred miles east of Mumbai in the Gadchiroli district, has experienced this firsthand. "I belong myself to the Scheduled Caste community," he told me. "I know the struggles."

Centuries of discrimination in India mean that people belonging to marginalized caste groups are disproportionately poor, and research by Jungari and others indicates that this inequality is a significant driver of caste-based disparities in health. On top of this, he told me, experiences of overt discrimination may additionally discourage people from engaging with health services. To take the example of maternal and infant health, which is the main focus of Jungari's research: in a 2021 review, he and his colleagues found that "lower caste" women were more likely to report experiencing mistreatment by health care providers compared with women belonging to other caste groups. This, combined with other barriers in accessing health services, may explain why women belonging to marginalized caste groups are more likely to have received no prenatal care during pregnancy and their infants are less likely to have received routine vaccinations compared with their counterparts belonging to other caste groups. "Discrimination is resulting in underutilization of health care," said Jungari.[163]

I was also curious to learn more about the origins of colorism in the Philippines, which was colonized and occupied by various countries throughout its history, including Spain, Britain, the US, and Japan. The Philippines was also significantly culturally connected to China through migration and trade, long before its colonization. All these factors have contributed to present day associations between lighter skin and higher social status, according to Rondilla. "When we're looking at colorism in the Philippines, what informs beauty standards is the multiple layers of colonialism," she explained to me. "Colorism in the Filipinx community is rooted in both anti-Black and anti-Indigenous racism," said Rondilla. However, she added, "our notion of lightness is not just about European-ness, it's also about being Chinese or Mixed Chinese or being East Asian, because, again, this has to do with who the Chinese have been historically in the Philippines and the type of power positions that they've held, and that also informs colorism for the Filipinx community."

In 2020, worldwide anti-racism protests put a spotlight on colorism in the marketing of cosmetics, particularly in the marketing of skin lighteners and whiteners to consumers predominantly in Africa and Asia. Amid this increased scrutiny, big players in the cosmetics industry, including Unilever, L'Oréal, Procter & Gamble, and Johnson & Johnson, announced changes to some of their products. For instance, Unilever said it would rename one of its skin creams, at the time called Fair & Lovely but since renamed to Glow & Lovely, and committed to removing references to "fair/fairness," "white/whitening," and "light/lightening" from its products' packs and communication, noting that the use of these words "suggest a singular ideal of beauty." L'Oréal and Procter & Gamble made similar commitments to remove references to certain words or rename products. Johnson & Johnson appeared to go a step further by discontinuing its Clean & Clear Fairness and Neutrogena Fine Fairness product lines, before replacing the latter with a new line named Neutrogena Bright Boost.[164]

To Rondilla, these rebranding efforts by companies equate to a repackaging of the same old problem. "I see those moves in the same way I would see someone putting a Band-Aid on cancer," she commented wryly. "They did it as a marketing move, they did it as this sort of political stunt," said Rondilla. "This is how these companies sort of pat themselves on the back thinking that they did a public good and, it's like, if you really wanted to do a public good, you would have just done away with this product."

In its June 2020 statement announcing its rebranding, Unilever said that "Fair & Lovely has never been, and is not, a skin bleaching product," adding that its products represented a "much-needed move from harmful chemicals like mercury and bleach, which consumers were using." But it isn't clear to me that mainstream products like Unilever's are *actually* moving consumers away from more dangerous, illegal products. Indeed, Grace's experience in the Philippines suggests the exact opposite: she had mentioned that seeing advertisements for mainstream skin lightening products was one of the motivating factors behind her dangerous attempt to lighten her skin, using what turned out to be a harmful, illegal knockoff. While colorism obviously wasn't invented by companies—and we are all collectively responsible for tackling this and other forms of racism in our societies—I am not convinced that their profiting from it is helping matters.

Beyond companies abandoning the promotion and sale of skin lightening products, Rondilla is also in favor of increasing media representation of darker-skinned people and educating people about the history of colorism. She is hopeful that colorism and its health harms can be addressed at a societal level, including by changing conversations within families. "I do feel optimistic about the future just because we talk about colorism way more now than five, ten years ago," she told me.

Speaking to Rondilla was part of what inspired me to start the conversation I had with my mother about colorism and our hair, which I found very healing. At one point during that discussion, my mom pointed out how much things had improved in her lifetime, comparing mine and my sister's experiences of colorism or featurism to hers, and even comparing her experiences to her mother's. My mom was a lot older than me or my sister when she first felt able to accept her natural, Afro hair, but this was still a huge leap compared to her mother (my biological grandmother from Ghana) who didn't see natural hair as acceptable for most of her life. "Whenever I got dressed and wanted to go out, my mother would look at my hair and say, am I going to do something about my hair?" My mum laughed, remembering when she first stopped chemically straightening her hair.

My grandmother came to live with us in the Netherlands when I was a teenager, after she was diagnosed with bone marrow cancer. She died from the illness in 2017, by which point I had left home and was living in the UK. Before my grandmother passed away, my mom said she had noticed a change in her relationship with her hair. "I think the first time I saw my mum just embracing her hair is after she had chemotherapy and she lost her hair, and it grew—and I saw some more acceptance of her natural look," my mom told me during our chat. "She looked beautiful, I mean, I remember always saying to her that she looked beautiful, even when she was bald."

Rondilla's research suggests that conversations within families are important in shaping people's perceptions of themselves and their self-esteem, both of which impact mental health and well-being. "When we look at colorism and mental health, one of the components that is often forgotten is intergenerational trauma," said Rondilla. "For me, engaging in the research was really about finding a sense of healing and trying to change a generational narrative," she explained.

It is positive to see improvements across generations, but the persistence of racial, ethnic, and caste health gaps suggests there is more work to be done to tackle racism and its health harms across societies globally. We have seen how racism in societies, including systemic and interpersonal forms of racism, are major contributors to health inequities. But the institutions of medicine and science aren't separate from these societies, nor are they immune to its ills; racism within these institutions—particularly in the Western world—also contributes to racial and ethnic health gaps globally. And that is what I want to examine next.

PART III

Racism in Medicine

CHAPTER 6

Racial Bias in Health Care

For many people, getting their wisdom teeth removed is a routine procedure. For Kanayo Dike-Oduah, a British Nigerian psychology teacher based in London, it became a near-death experience.

It isn't uncommon to experience some pain for a couple of weeks following wisdom tooth removal. But in August 2019, Kanayo was still in pain—three months after her procedure. She went to see her dentist, who told her he couldn't see any sign of an infection and that it might be nerve pain that she was experiencing. The dentist suggested that she go to see her family doctor or GP, who prescribed her carbamazepine—a drug to treat the suspected nerve pain.

After two weeks of taking the drug as instructed and seeing no improvement, Kanayo went back to her GP. "I said, well, it's not relieving the pain that I was experiencing, and, in fact, I've noticed that my shins, so my legs, have gotten scaly, like dry skin, which is unusual for me, and that I was also developing bumps on my tongue as well," she recalled. "Little did I know that these were listed side effects of the drug," Kanayo told me.

The GP advised that she increase the dose of the drug, doubling it to two tablets twice a day. However, within a week of doubling her dose of carbamazepine, Kanayo's health suddenly took a turn for the worse. "All of a sudden I started to deteriorate," she said. "It first started off looking like I just had puffy eyes, especially just beneath my eyes.

And I went to A&E [Accident & Emergency] because it felt like my breathing was a bit labored," she remembered.

At the A&E department, Kanayo was told she appeared to be having an allergic reaction, probably to food. "I told them that I'd started taking this new drug," said Kanayo. Being an avid note taker and someone who is very diligent about her health, she pulled up some notes she had made about the medicine. "I said, this is the dose, this is when I've been taking it, this is when I started it," she recalled. Yet no connection was made between the drug and her symptoms. She was given an antihistamine—a type of medication often used to relieve allergy symptoms—and sent home.

The following day, Kanayo was back at A&E. Her symptoms had already worsened. "It was so rapid in terms of how fast I was deteriorating," she told me. Her face had become even more swollen and there was pus coming out of her eyes. Her lips were also cracked and swollen, and she had developed rashes all over her body. "I went back to A&E and they said, 'Oh, it looks like you've got tonsilitis and conjunctivitis,' and, indeed, my tonsils were swollen and ulcerated," said Kanayo. "But there was nothing to explain my cracked lips, nothing to explain the puffiness underneath my eyes and the rashes on my skin," she noted. "Again, I was just given medication and sent off on my way."

A day later, Kanayo felt so ill that she couldn't even bring herself to go back to A&E. "Literally my whole body was in pain," she said. "The capillaries in my palms had burst, so my palms were literally displaying a rash as well," she added. "We had to call the paramedics," said Kanayo. "They came to see me, and by this point you could clearly see that I did not look like my usual self," she emphasized. "I had to take out my extensions, because my scalp literally felt like it was on fire." The paramedics didn't seem to recognize this, though. "Instead of rushing me to hospital, they gave me paracetamol that I actually wasn't even able to swallow, because my mouth was so ulcerated, and they went on their way," she said frustratedly.

This is when Kanayo decided to start documenting what was happening to her. "I love taking photos. I love selfie gang," she told me, laughing for a moment. The photographs of her face aren't funny at all, though. Looking at them all I could see was a woman around my age, who looked extremely unwell and like she was in a great deal of pain. "Even though I was looking terrible, I just thought, well, I have to keep on taking photos in case I die," she said solemnly. "In case I die, I need these pictures to show what happened to me."

The pictures saved Kanayo's life. After the paramedics left, she was in so much discomfort that she couldn't sleep. She could feel that her breathing was labored and, concerned, she decided to share the photographs she had taken with her sister. "I sent photos of myself to my younger sister and I said to her, 'Can you just pop into any pharmacy and just show them what I look like?' and, 'These are all the medications that I'm on,'" she said. At this point, Kanayo was taking a cocktail of different medications including the original drug, carbamazepine, as well as steroids, antihistamines, and paracetamol, which she was struggling to swallow. Her sister explained the situation to a pharmacist, who, after seeing the photographs of Kanayo's face, advised that she seek urgent medical attention.

"This person looks like they're having a drug reaction, you need to get them to a hospital now," the pharmacist told her sister. Kanayo was rushed to the hospital by her family. She collapsed the moment she walked through the A&E doors. "My body just gave up and they had to resuscitate me," she told me. Kanayo was admitted to the hospital and diagnosed with Stevens-Johnson syndrome, a rare, life-threatening disorder that can be caused by an adverse reaction to certain medications. Carbamazepine—the drug Kanayo had been prescribed a few weeks earlier—is listed on the National Health Service (NHS) website as one of the medicines that most commonly causes the syndrome. "Externally I was wasting away, but internally, the lining of my organs were also gradually being destroyed," Kanayo explained. "That's what caused me to—literally, my body just went into cardiac arrest," she said.

After Kanayo regained consciousness in her hospital bed, she was seen by a physician. "A Black consultant was the one who saw me," she recalled. "When I came round she said, 'This is not what you normally look like, you're in a lot of pain.' And I said, 'You're the first person to actually acknowledge that,'" Kanayo remembered. "Every other point of care—and unfortunately they were White—every other point of care, it was just like, 'Yeah, just take paracetamol' and, 'Sometimes it gets worse before it gets better.' There was no acknowledgment of the fact that I was in severe pain and that was not what I normally looked like," she stressed. It was more than a week until Kanayo was well enough to be discharged from hospital. Reflecting on the experience, she said: "I just thought to myself, if I was presenting all of this, but my skin was just lighter—if I was a White woman—then what would the response have been?"

Like me with my eczema, Kanayo suspects that her condition might have been diagnosed more quickly if she were White. But while the delay in diagnosis was a minor inconvenience in my case, in Kanayo's it was a life and death matter. And while it is impossible to definitively determine the extent to which her skin color influenced how she was treated, there is plenty of evidence that the systemic and interpersonal racism present within societies across the globe also permeates medical institutions. I wanted to investigate this—starting with taking a closer look at how racial bias within the health care workforce can exacerbate existing racial and ethnic health gaps.

Following her terrifying ordeal, Kanayo became curious about the issue of racial bias in medicine. She wanted to learn more about the rare condition she had experienced, so she visited the NHS website and typed "Stevens-Johnson syndrome" into its search function. "Lo and behold, all you can see is what this syndrome looks like on White skin," she told me. I performed the same search several months after my conversation with her and, indeed, there was only one photograph on the page, showing reddish sores on a pale-skinned arm.* Out of curiosity, I then tried typing "Stevens-Johnson syndrome" into Google Images. I had to scroll down a bit before I found any photographs showing the condition on darker skin, and one of those photographs was of Kanayo. I clicked on it, and it led me to a scientific article titled "Stevens-Johnson Syndrome in a Patient of Color: A Case Report and an Assessment of Diversity in Medical Education Resources." It was a medical case report about Kanayo herself, authored by a group of mainly medical students, including lead author Darlene Diep, a medical student at the Burrell College of Osteopathic Medicine at New Mexico State University.[165]

Case reports provide a way for medical students and doctors to learn from individual patient experiences. I found it really touching to see that Kanayo had shared her photographs in an effort to turn her awful experience into an educational resource for the international medical community. "Thanks be to God that now the pictures are able to contribute toward medical education," she told me.

The case report identifies Kanayo as a "25-year-old Black woman, with Fitzpatrick skin type V (brown skin that rarely burns and tans very easily)" and describes the symptoms she initially presented with, before noting that "multiple misdiagnoses and the eventual exacerbation of the patient's condition culminated in her hospital admission, where she collapsed and was resuscitated." It goes on to state that "the nature of the disease progression in this case prompted our investigation into the lack of representation of skin of color in current medical training resources regarding SJS [Stevens-Johnson syndrome]," concluding that

* The NHS has since updated its webpage on Stevens-Johnson syndrome to include a photograph of a darker-skinned patient with the condition. As of 2023, this photograph appeared relatively high up in a Google Image search of Stevens-Johnson syndrome. I think this is a great example of how a relatively small change can potentially have a large impact.

"our assessment demonstrates that there is a significant underrepresentation of SJS in skin of color in medical educational resources," and that "increased inclusivity of skin disorders in patients of color is crucial in training health care professionals to recognize life-threatening cutaneous disorders quickly and accurately in such patients." Specifically, they found that among five commonly used medical education resources, 80 percent of the images showing Stevens-Johnson syndrome were of people with Fitzpatrick skin types I–III (lighter skin tones), while just 20 percent showed people with Fitzpatrick skin types IV–VI (darker skin tones).

These results echo findings from an earlier study, which we touched on at the start of the book, led by Patricia Louie—the sociologist at the University of Washington. Louie's study compared the representation of different skin tones in widely used medical literature with that of the US population. A 2012 survey, which asked people to define their skin tone according to the Massey-Martin skin color scale, estimated that 63 percent of the US population has a "light," 26 percent has a "medium," and 11 percent has a "dark" skin tone. Yet Louie's analysis of images in the medical literature using the same scale found that 75 percent of images showcased "light" skin, while skin classified as "medium" or "dark" accounted for just 21 percent and 5 percent of images, respectively. The dominance of Western medicine in much of the world means that this bias in medical education is harming darker-skinned people globally, even in regions such as Africa where darker skin tones are common.[166]

The case report by Diep and colleagues references another case chillingly similar to Kanayo's, published in a separate report two years earlier, of a "46-year-old African American female patient" who visited a US emergency room three times before being admitted and diagnosed with Stevens-Johnson syndrome. "The similarities to our patient's case underscore a trend in terms of the lack of early recognition of this life-threatening condition, especially in patients of color," Diep and colleagues wrote in their report. Case reports are anecdotal evidence, of course, but more systematic research also points to a problem in the ability of doctors and other health care professionals to recognize symptoms on darker skin. A survey of six hundred users of an international online medical teaching platform found that only 5 percent said they felt confident diagnosing skin conditions across a range of skin tones, for instance. The results of that survey also showed that health care professionals who reported having been exposed to more inclusive resources during their training tended to report greater confidence in their ability to diagnose conditions across diverse skin tones. Confidence was also associated with geographical location—it was higher in Africa and Latin America, although still only 29 percent and 11 percent of doctors in these respective regions reported feeling confident. It varied by specialty too; confidence was highest among dermatologists (unsurprisingly, since they specialize in the diagnosis and treatment of skin disorders) and lowest among pediatricians, emergency medicine, and pediatric emergency medicine specialists.[167]

I have always felt extremely fortunate to live in a part of the world where I know that if I become seriously unwell, an ambulance will be able to come and get me and hopefully bring me to a well-resourced hospital to receive care. But speaking to Kanayo and learning more about biases in medical diagnosis left me feeling less safe. I wanted to speak to a doctor, in the hope that they might offer some sort of reassurance for me and other people with darker skin tones, especially those of us living in predominantly White countries. Carlos Charles very much fit the bill. He is a New York–based, board-certified dermatologist who in 2012 founded his own dermatology practice called Derma Di Colore, specializing in providing care for patients with darker skin tones. "I saw a need at the time," Charles told me. It is a need that remains; I could sense it from Charles's enthusiasm to talk to me and his excitement about the subject of my book. "There's a huge need," he emphasized. "All of these major textbooks in dermatology, when you look through them, most of the photos, most of the images are in lighter skin. That doesn't represent what the world actually looks like," he said. "I think that's a huge problem."

Charles was motivated to do more to help communities that, in his words, looked like him. In his field of medicine—dermatology—the ability to recognize signs and symptoms on a range of skin tones is particularly important. "We diagnose things on the skin, but one of the beautiful things about our field is that a lot of the diagnoses that we make are speaking of something that's going on internally, which is what was attractive to me in dermatology," Charles explained. "So, it's kind of our job to be gatekeepers to catch things that may be going on inside," he said, pointing out that dermatologists often need to pick up on conditions that present with symptoms on the skin but that may require care from other specialists, such as rheumatologists or internists. "If we miss these things early on, it's going to lead to a poor health outcome, because whatever condition it is, may become worse," he said.

Charles is also passionate about empowering his patients with information and dispelling harmful medical myths, like the idea that darker-skinned people can't get skin cancer. "Skin cancer is less common in darker skin overall, both non-melanoma and melanoma skin cancer, but it does occur," he said. "In darker skin it tends to be diagnosed much later," he added, at which point treatment is more difficult. A 2016 study in the US found that among almost ninety-six thousand patients with melanoma, the deadliest form of skin cancer, White patients had the longest survival time while Black patients had the shortest. The disease was also more advanced among Black patients compared to White patients on average, consistent with diagnosis occurring at a later stage. Charles thinks this delay in diagnosis may be partly due to some physicians struggling to recognize the condition early enough among darker-skinned people, but he thinks widespread myths about darker skin not requiring protection from the sun may also contribute.[168]

I have certainly come across these myths. As I was talking to Charles, I thought back to the time one of my schoolteachers advised my classmates to pack sunscreen ahead of a planned school trip and then looked over to me and laughed, pointing out that sunscreen *obviously* wouldn't be necessary for me, since my brown skin provided me with natural protection from UV (ironically, I ended up experiencing my first ever sunburn on that trip). Myths like these may also be perpetuated by lack of skin tone diversity in sunscreen ads or public health messaging, and they can impact health behaviors starting from a young age.

A survey in the US, involving more than five thousand fifth graders—schoolchildren aged 10 or 11—and their primary caregivers, found that Black and Hispanic children were less likely than White children to almost always use sunscreen. This pattern persists into adulthood; a separate survey of more than 2,000 US adults showed that just 39 percent of African American respondents said they ever used sunscreen. The fact that many sunscreen formulations cater mainly to lighter skin tones probably doesn't help matters. In a blog post published by 4.5.6 Skin, a skin care company cofounded by Charles that specializes in making products for darker skin tones, Charles points out that a lot of sunscreen formulations leave a visible white residue or "white cast" when applied on darker skin, which can put people off using sunscreens altogether. The blog post encourages people of all skin tones to use sunscreen and goes on to suggest some examples of formulations that may be more suitable for people with darker skin.[169]

While preventative measures such as using sunscreen benefit people's health and should be encouraged, this doesn't excuse doctors and other health care workers from being able to care for patients across all skin tones—especially in the field of dermatology. Charles believes that all dermatologists, regardless of their skin color, can learn to take care of patients with darker skin. "Diseases in dark skin can look very different than they do in light skin," he noted, when I told him about the difficulties I faced in getting diagnosed with atopic dermatitis or eczema.

"Atopic dermatitis in darker skin can look very different than it does in lighter skin," he explained. "You have to know what to look for." But that's not all, according to Charles: it is also crucial that doctors learn how to communicate important information to patients, taking skin color into consideration.

As I was writing this chapter, I had a phone call with my younger sister, who lives in the UK and who also experiences eczema. She mentioned feeling frustrated during a telephone consultation with a doctor, who advised her to look out for "reddening of the skin" as a sign that her condition was worsening and might warrant a referral to a specialist. "My skin doesn't turn red," my sister recalled telling the doctor, who failed to suggest any alternative signs that she could look out for on her dark brown skin. My sister told me it wasn't the first time she'd had an experience like that when engaging with health care. As Charles told me, "trust is such an important part of that dynamic."

For Kanayo, that trust has been shattered. After she told me about everything she had been through, I asked her how her experience had affected her trust in the health care system in England, where she lives. "I know that I have to use my voice more than I ought to," she began. "I guess when you are in pain, sometimes you fall into that place where you think, well people should just know that I'm in pain. But as a Black woman—we literally have to vocalize our pain. If we can't vocalize it, then we need to have people alongside us who can vocalize it for us," she said. Kanayo was already diligent about her health before her ordeal, but she has become even more so since, to the point of second-guessing everything that she is told when she goes to the doctor. "When the doctor tells me something, I make notes, I research it myself, I use Google scholar—I will read the medical trials that have been done on so and so drug that they're trying to prescribe me, just to give me that peace of mind," she told me.

I asked Charles what he thought about Kanayo's approach, and what advice he would offer to patients who, like her, feel a particularly strong need to advocate for themselves and to double-check everything they are being told in health care settings. "I don't think it's incumbent upon the patient to Google and figure it out," Charles replied. Indeed, I know from my own experience that trying to self-diagnose on the internet can be stressful at best and dangerous at worst: not to mention that it also implies additional labor for patients of color. At the same time, I relate to Kanayo's urge to try to protect herself from racial bias at the doctor's office. Aside from magically eliminating racism from the world, Charles suggested: "Doing some research in terms of who you're seeing can be helpful."

Unfortunately, patients don't always have a choice about who they see. "I see patients mostly that have health insurance," Charles mentioned, adding that this can present a barrier for both uninsured and insured patients. "I can go on and on about the challenges of insurance," he said. Even in the UK, which has a system of universal health care, the amount of money you have can influence your care options as a patient. I forked out on two separate occasions to see private dermatologists for my own skin problems in the hope that they might bring me closer to some kind of diagnosis or solution (they didn't). "The reality is, every human being has got a bias," Kanayo told me. "And it's just so important that we check our biases," she said.

I agree with her—physicians should regard checking their conscious and unconscious biases as a central part of abiding by their Hippocratic oath. Because interpersonal racism that manifests in doctor-patient interactions impacts health.

During my time as a PhD student in Oxford, I had an unsettling encounter at what was otherwise a fairly routine visit to the GP. I don't recall exactly when it was or why I had come to see the doctor on this particular occasion, although I suspect it may have been to get a prescription for a contraceptive pill. I do remember that during the appointment, the doctor started asking me lots of questions about my sexual history, and she appeared mistrustful of my answers. After a while, she recommended that I should get tested for HIV and gave me information about a clinic where I could go to get tested. I didn't think much of this, until I was standing up to leave.

I had come to the appointment straight from my work at the laboratory and as I stood up and leaned forward to pick up my jacket, the lanyard I was wearing around my neck swung forward slightly. Attached to the lanyard were two cards—one with an NHS logo on it, because the laboratory where I conducted my PhD research was based in an institute attached to a hospital, and another with "University of Oxford" printed on it in capital letters. The doctor glanced at my lanyard and suddenly I felt her whole attitude toward me shift.

"Oh, do you work at the hospital?" she asked.

"I work in the research institute nearby. I'm a PhD student," I answered.

"Wow, really? At the university?" she replied, referring to the University of Oxford.

"Yeah," I told her, and she seemed impressed.

She asked me a bit about my research on viruses and we continued chatting as I was putting on my jacket and walking toward the door. Before I opened the door to exit, she said something along the lines of: "By the way, don't worry about getting an HIV test. I think it's very unlikely that you have HIV." She said it so casually that I didn't really process how problematic it was until I had walked out of the door.

As ever with these sorts of interactions, I don't have any proof that the doctor did anything wrong and, in fact, I don't believe she had any bad intentions. Also, I wasn't exactly harmed by the experience; at the time I shrugged it off as a microaggression. I remember complaining about it briefly to my mom and sister, and to a few of my friends who could relate, but that was pretty much it. I wasn't even planning to write about it in this book. But somewhere during the process of listening to other people's experiences of health care and starting to write, I began to realize that what happened to me was part of a much larger problem of bias and stereotyping in medicine.

Reflecting on the decisions that the doctor made when she saw me, I can now see that they were probably driven by a number of unconscious biases. From the moment I entered her office and sat down, she will have started making unconscious assumptions about me based solely on my appearance. For instance, she will have racialized me as Black and as she began asking me questions about my sexual history, she may have been influenced by stereotypes about Black women being hypersexual or promiscuous. At the same time, without being aware of it, she may also have been influenced by harmful stigma and stereotypes about people living with HIV, including negative stereotypes connected to promiscuity and notions of immorality.

When she learned that I was a PhD student at an elite university, this may have subconsciously contradicted some of her prior assumptions about me as a Black British woman—including about my social class (in addition to racism, classism is also alive and well in the UK). All of a sudden, I didn't fully match her unconscious idea of what an HIV-positive person should be like and so her assessment of my risk changed.

This is extremely problematic. In addition to being racist and classist, it also perpetuates stigma against people living with HIV as well as those who choose to get tested—something we should be strongly encouraging. When I relayed my experience to Saidy Brown, the South African HIV activist, she agreed. "People are very judgmental towards people who are living with HIV, just by virtue of them living with HIV," Saidy told me. "There is a lot of racism and there is a lot of stigma," she said.

I also shared my experience with Jay Kaufman, the epidemiologist at McGill University in Canada whom we met earlier. He highlighted the fact that doctors and other health care workers are subject to all the same biases and exposed to all the same harmful racist stereotypes as the rest of us. "Doctors go home, and they watch the same TV shows that everybody else watches, they get exposed to this just like everybody else," Kaufman told me. Recognizing and acknowledging this is empowering for doctors and patients alike.

I have an enormous amount of respect for people who work in the medical profession—and included in that respect is an understanding that they are human beings who are fallible. That extends to the doctor who saw me in Oxford; my intention in sharing that story was not to criticize her as an individual but rather to ask what factors contribute to these sorts of behaviors or practices in medicine, and how can we learn from them to make the institution of medicine fairer. This isn't a new idea: many people both within and outside of the medical field, including plenty of medical students and doctors across the globe, have long been working to try to make medicine more equitable.

Malone Mukwende, the medical student in London and coauthor of *Mind the Gap,* is also the founder of Black and Brown Skin, an online platform that showcases images of various health conditions as they appear on darker skin tones, and of Hutano, a social health platform for Black and Brown people. In a Talks at Google session in 2021, Mukwende laid out his vision for the future of medicine. He pointed to medical education as a key area for improvement. "I think the next step in UK [universities], especially in the health care space, is making sure that students—wherever they are on their individual courses—continue to flag up when they find that, actually, we've been taught about meningitis but there were no pictures on darker skin," he said. "Eventually, I think we'll start to see an overhaul or big shift," Mukwende went on. Such a shift will necessitate engagement and commitment not only from medical students, but from across the medical profession. "As an entire profession, I feel like we haven't done enough as of yet," he concluded.

Fortunately, Mukwende isn't alone in his desire and effort to improve things. In addition to resources like *Mind the Gap* and Black and Brown Skin, there are other tools and resources that are being developed to help make medical education more representative of humanity. "There's a lovely doctor, her name's Dr. Sharon Belmo, and she, in fact, has started rewriting the curriculum for dermatology students," Kanayo told me. Belmo is a dermatologist herself and in 2021 she wrote the first "Dermatology for Skin of Colour" syllabus for the UK dermatology specialist training curriculum. She is also the cofounder and clinical lead of the Centre of Evidence Based Dermatology Skin of Colour Resource, launched in 2016. In the US, there have also been efforts to increase skin tone diversity in medical education. For instance, medical students and faculty at the University of New Mexico's School of Medicine have established a Visual Atlas of skin conditions, which collects and features images of dermatological conditions on a range of skin tones.[170]

Companies are increasingly recognizing the demand for better educational resources too. US firm VisualDx, which makes a clinical decision support system by the same name, boasts on its website that almost 29 percent of the pictures in its medical image resource are of Fitzpatrick skin types IV–VI. But tackling bias in medical education is not as simple as just adding more images of health conditions on darker skin. While medical textbooks in the US tend to showcase conditions on mainly lighter skin tones, there is a notable exception to this: sexually transmitted infections (STIs) are more often shown on darker skin tones, which may perpetuate racist stigmatization and stereotyping (I suspect this sort of bias might have influenced the doctor I saw in Oxford).

A 2019 US study, which reviewed images from two commonly used medical textbooks and one frequently used teaching set, found that the proportion of images showing skin of color was higher for STIs compared to other types of infections (for images of STIs the proportion was estimated to be between 47 and 58 percent, while for images of other infections it was estimated to be 28 percent). When it comes to medical education, skin tone biases in any direction are unhelpful both for those learning to recognize and diagnose conditions, and for their prospective patients. Teaching would-be doctors how health conditions appear on diverse skin tones, on the other hand, stands to benefit everyone.[171]

Unfortunately, a lack of diversity in medical textbooks is just one example of how racism in medicine manifests. Conscious and unconscious biases are also evident in the extent to which people are listened to or believed when they attempt to seek health care. These biases often affect people along gender lines as well as race or ethnicity. As I was gathering stories for this book, I noticed that the experience of not being listened to seemed to be a common thread in people's negative encounters with health care, particularly among women of color.

As we have seen, Serena Williams wasn't listened to when she asked to be examined and treated for suspected blood clots in her lungs after she gave birth, for instance. "Doctors aren't listening to us," were her exact words, when she spoke to the media about what she went through.

D'lissa Parkes wasn't listened to when she was sent home from the hospital a day before she died. Instead, according to her mother, Sylvia, the doctor she saw made inaccurate assertions about Black women's pelvises being shaped differently to White women's.

Naomi Williams wasn't listened to during the eighteen visits she made to the hospital in the months leading up to her death, despite her mother, Sharon, sending a written complaint. "The Hospital continued to treat Naomi in an uncaring way whatever we did or said to them," said Sharon.

Carol Ighofose wasn't listened to when she told the paramedic who was attending to her, "I am a doctor, I do believe I'm having a heart attack," and was told in response, "Oh, you doctors always think of the worst."

And Kanayo Dike-Oduah wasn't listened to when she visited the emergency room *three times* before finally collapsing and being admitted to the hospital for what turned out to be a life-threatening condition.

These cases are just a few of the examples that we know about; cases where the people involved or their families decided, bravely, to share their experiences publicly. There are probably a lot more stories like theirs, including many more cases where people's lives were similarly endangered and, in some cases, even lost. Listening is important in all health care interactions, not just during medical emergencies. Good communication between doctors and patients is associated with better health outcomes overall. Yet, unfortunately, racism—including conscious and unconscious bias—often gets in the way of this.[172]

One of the most pernicious things about racism is not just that it is often difficult to prove, but that it is those who experience racism who are most often tasked with providing said proof. This is also true of racism in medicine and is probably all the more reason why Susan Moore, a fifty-two-year-old Black family doctor in the US who died from Covid-19 in December 2020, decided to document her treatment after being admitted to an Indiana hospital with the disease. Susan shared her experience in a heartbreaking video, which she posted on social media about two weeks before she died. In the video, recorded on her smartphone in selfie mode, Susan is lying in a hospital bed with oxygen tubes clinging to her nostrils. Her voice sounds hoarse and her breathing is clearly labored. She describes being in excruciating pain, yet being denied pain medication until a CT scan revealed just how ill she actually was. "I don't feel comfortable giving you any more narcotics," she remembered a doctor telling her. "I was in so much pain from my neck, my neck hurt so bad. I was crushed. He made me feel like I was a drug addict, and he knew I was a physician," said Susan in her video. "I don't take narcotics. I was hurting," she added.

Instead of treating her pain, the doctor wanted to send her home. According to Susan, he threatened to discharge her on Saturday night at 10:00 P.M., in the dark. "This is how Black people get killed," she said. "I had to talk to somebody, maybe the media, somebody, to let people know how I'm being treated up in this place."

Exasperated, Susan said she had started to ask whether she could be transferred to another hospital for treatment. "Next thing I know, I'm getting a stat CT of my neck, with and without contrast. The CT went down a little bit into my lungs and you could see new pulmonary infiltrates, new lymphadenopathy, all throughout my neck. And all of a sudden, 'Yes, we'll treat your pain,'" she recalled. "You have to show proof that you have something wrong with you, in order for you to get the medicine," she continued. "I put forward and I maintain, if I was White, I wouldn't have to go through that," she concluded.

Susan was eventually discharged from the hospital but within twelve hours she was readmitted to a different hospital, where she died on December 20, 2020. She was survived by her nineteen-year-old son, Henry Muhammad. According to Henry, his mother had gotten used to having to advocate for herself in order to get decent medical care. "Nearly every time she went to the hospital she had to advocate for herself, fight for something in some way, shape or form, just to get baseline, proper care," he told the *New York Times*, following her death. Henry didn't get a chance to say goodbye to his mother before she died. He told ABC News: "I am outraged beyond words . . . because if what my mom thinks was true and that it was racism, and they neglected her because of that, nobody should go through that."[173]

Henry's outrage was echoed around the world; Susan's case received widespread media coverage and sparked conversations about racism, including within the medical community. "If a physician can't be heard by her own peers to save her life, then who will listen?" asked an op-ed in the *Washington Post* by four health experts in the wake of Susan's death. The authors, Aletha Maybank, Camara Phyllis Jones, Uché Blackstock, and Joia Crear Perry, all based in the US, wrote: "The deaths of Mr. George Floyd and so many others mistreated, injured or killed at the hands of our policing system have made us accustomed to seeing the video. But injustice in health care is rarely broadcast from cellphone videos or shared for thousands to witness. This injustice often remains invisible to the public—unless, of course you are a member of the community experiencing it."[174]

The injustice certainly isn't invisible when you look at the statistics. By the time of Susan's death in 2020, it had become abundantly clear that there were vast racial inequalities in who was getting ill and dying from Covid-19, both in the US and elsewhere in the world. As a journalist reporting on these issues, I noticed that there was growing mainstream recognition that racism could be contributing to these disparities. But Susan's death sparked a more specific conversation, not only about the role of systemic racism within societies, but about the role of interpersonal racism within the medical profession, in influencing people's health outcomes.

Susan was in pain, and despite the fact that she was a physician herself, the doctor who attended to her wouldn't take her word for it. This isn't an anomaly: an analysis of data on acute pain management in US emergency rooms from 1990 all the way up to 2018 found that Black and Hispanic patients were 40 and 25 percent less likely than White patients to be given pain relief. These inequalities also exist outside the US and extend to chronic as well as acute pain. A small retrospective study of people with Parkinson's disease in the UK found that significantly fewer people of Black or Asian ethnicity compared to White people reported receiving prescribed pain relief, despite all reporting a similar pain burden in a questionnaire. Harmful stereotypes about Black skin being thicker than White skin or Black people being less sensitive to pain are pervasive in medicine and may partly contribute to racial and ethnic gaps in pain relief.[175]

As is so often the case when it comes to notions of biological difference between racial groups, ideas about racial differences in pain perception and tolerance can be traced back to pseudoscience aimed at justifying the subjugation and enslavement of Black people. In the southern US in the 1800s, the doctor and plantation owner Thomas Hamilton conducted a series of horrifying experiments with the goal of demonstrating that Black people have thicker skin than White people. Believing that Black enslaved people possessed some unique ability to withstand pain may have helped to ease the consciences of their enslavers. It was to this end that John Brown, an enslaved Black man on a plantation in the state of Georgia, was loaned to Hamilton as an experimental subject. Brown later escaped to England and published an autobiography, in which he described the suffering that Hamilton had inflicted upon him. Hamilton applied "blisters to my hands, legs and feet, which bear the scars to this day," Brown wrote. "He used to blister me at intervals of about two weeks." This continued for nine months, until "the Doctor's experiments had so reduced me that I was useless in the field," he reported.[176]

While I highly doubt that doctors nowadays are deliberately intending to perpetuate this legacy of harm, it has become clear to me in writing this book that many historic, racist ideas have been passed on to generation after generation of medical students without enough critical reflection on their origins or scientific validity. That is a huge problem. The decision to prescribe pain medication (or not) represents just one among a wide array of decisions that doctors and other health care workers are faced with on a daily basis, often under enormous pressure. It isn't unreasonable to therefore assume that racial bias might influence these other medical decisions too, including potentially life-altering ones.

Consider, for example, the decision on whether it is necessary to amputate a limb. In the US, Black people undergo diabetes-related amputations at almost three times the rate of non-Black people, and a number of studies suggest that racial disparities in amputations remain even after accounting for factors such as socioeconomic status, health insurance coverage, and preexisting health. If there is any chance whatsoever that conscious or unconscious racial bias might be responsible for even a tiny fraction of those—and the countless other racial and ethnic health gaps that we've come across so far in this book—then I think it is worth having a conversation about at the very least.[177]

Olamide Dada, a junior doctor based in the UK, hasn't been shy about starting that conversation. She is the founder and chief executive of Melanin Medics, a nonprofit organization that supports aspiring and current doctors of African and Caribbean heritage in the UK. I caught up with her toward the end of 2021, and she shared her perspective on racism in medicine as someone working within the profession. Speaking with Dada gave me a lot of hope; she had graduated from medical school that summer and very much embodied a new generation of doctor. She is passionate about equity in health care and about making hospitals more welcoming places for both patients and health care workers of all backgrounds. To achieve this, she understands that it is necessary to have uncomfortable conversations about racism and the role it plays in patient–health care worker interactions, as well as in interactions between health care workers.

"The disparities that we see in health care shouldn't surprise us when we consider the disparities that we see in the health care workforce," Dada told me. "The way that we see colleagues being treated, let alone by other colleagues—how do we think patients are treated when it's a lot more of a vertical relationship? When patients view health care professionals to know more, to be the experts, to really be the ones responsible for caring for them? Oftentimes, people feel like they can't challenge the decisions of health care professionals," she explained. "I think a lot of work needs to be done, not just making hospitals a more equitable place and a more welcoming place for communities of all races, but also considering, how are we actually treating the staff? Are we empowering members of staff to go back into their communities and to empower their communities?"

Dada launched Melanin Medics in 2017, while she was still a medical student. It started out as a blog aiming to increase representation of African and Caribbean people within medicine in the UK, but quickly grew into an organization providing everything from networking events to mentoring opportunities for aspiring and current doctors from African and Caribbean backgrounds. Dada's motivation to start the blog was connected to her own personal journey into medicine: she wished she'd had more support available to her when she was applying to medical school herself.

"There are many challenges that I faced. One was growing up in an area that notoriously didn't really send students to medical school—or to university, let alone medical school," Dada told me. "And even aside from that, once I had decided that medicine is what I wanted to do, and I had continued to achieve good grades, then my chemistry teacher didn't want to give me the predicted grades that I would need to apply for medicine in the first place," she said. "The ironic thing was, at the end of the day, the grade she didn't want to give me, I ended up achieving," she added, triumphantly.

Halfway through her application process, Dada found herself a mentor. "She was so influential. She was so, so inspiring. She grew up in the same area as me, she went to the same school as me," said Dada. After being accepted into medical school, Dada began to realize that the challenges she had faced during the application process were shared by some of her peers. "I just realized that so many other Black students were going through similar things where their teachers didn't want to give them the right predicted grades, so under-predicting the grade that they think they would go on to achieve, where people didn't have access to mentors," she explained. "I really wanted to be part of the solution. I really wanted to just show people that you can come from this background and succeed and thrive in the medical space."

Dada believes that increasing diversity within the medical profession will help to mitigate racial bias and reduce racial and ethnic inequalities in health. In England, where she is based, data suggest that the medical workforce is becoming increasingly diverse. People of Black and Asian ethnicity are overrepresented among National Health Service (NHS) staff compared to their share of the country's working-age population, and the number of doctors from a Black or other ethnic minority background increased by 21 percent from 2017 to 2020. But these statistics mask persisting inequalities. Black people in particular remain underrepresented among consultants—the most senior doctors in the NHS—and people from Black and other ethnic minority backgrounds are underrepresented in clinical director and academic roles. Black and other ethnic minority doctors in England additionally report experiencing greater levels of discrimination and harassment at work, compared to their White colleagues.[178]

"I'm very lucky that I haven't had to deal with overt racism as much as some of my colleagues. But that doesn't mean it doesn't exist, just because you don't see it around you, just because you're fortunate enough to work with people who value you, who value your contribution to the workplace, doesn't mean that your other colleagues aren't going through much worse experiences," said Dada. "I think we just have a long way to go in terms of just making sure that the health care space is welcoming for people from diverse backgrounds," she added. Improving representation in medicine at all levels will be important for achieving that—and it stands to benefit patients as well as health care workers. Research from the US, where Black people account for 13 percent of the overall population but just 5 percent of active physicians, suggests that increased representation translates into tangible health benefits for patients. We have seen how the infant mortality rate among Black babies is more than twice that among White babies, but a 2020 study found that this higher mortality rate was halved among Black infants who were cared for by Black physicians. This echoes an earlier study conducted in Oakland, California, which found that Black men who were randomly assigned to Black doctors had better health outcomes than those assigned to a non-Black physician. The researchers behind that study estimated that Black doctors could reduce the Black-White gap in cardiovascular mortality by 19 percent among men.[179]

"It's so important to see people that look like you in whatever space that you find yourself. Particularly for patients, it can be quite encouraging; I know with my Black patients, you can see the way that their demeanor changes when they see me," Dada commented. "You can just feel the difference and that difference is them feeling more comfortable," she told me. Based on my own personal experiences as a patient, I suspect that the real benefit here may be derived not from having a doctor that is necessarily the same race or ethnicity as you, but rather from having a doctor whom you feel you can trust. After all, health care workers of all races and ethnicities can perpetuate racism (especially when their medical education is racially biased). Trust is key, and a lack of it—for instance due to past experiences of racism—can discourage people from seeking medical care altogether, even when they most urgently need it. Dada hopes that increasing representation across the medical workforce will go some way toward rebuilding trust among marginalized groups.

But in order to maintain that trust and fully empower patients, it will also be necessary to address biases that are deeply embedded in the very core of Western medicine—to systematically reckon with the racist legacies that persist in medical practices, guidelines, and thinking across the globe.

CHAPTER 7

Race-Based Medicine

Kevin Henry was born and grew up in the US in the small, majority African American town of Mound Bayou, Mississippi. He attended John F. Kennedy Memorial High School in the 1980s before leaving to study at Mississippi State University, where he played American football. Kevin's talent saw him drafted into the National Football League (NFL) in 1993 by the Pittsburgh Steelers, where he played as a defensive lineman for eight seasons. In 1995, the Steelers won the American Football Conference championship and advanced to the Super Bowl, where they were defeated by the Dallas Cowboys. But in 2001, at the age of thirty-three, Kevin's sports career came to an end.

Like many other American football players, Kevin had begun struggling with what he suspected were the long-term effects of the numerous concussions he had sustained while playing over the years. "Football doesn't give you an expiration date—you just expire," Kevin said during an interview on the ABC News show *Nightline* in February 2021. He said he experiences frequent headaches, depression, and memory loss, which cumulatively leave him unable to work. In the years since Kevin first started playing professional football, multiple studies have emerged demonstrating a link between head injuries in the sport and long-term conditions affecting the brain, including chronic traumatic encephalopathy and dementia.[180]

In 2013, the NFL reached a financial settlement with former players who had sued over head injuries sustained while playing, agreeing to pay compensation to players who received certain diagnoses, including dementia. Kevin decided to apply for compensation through the settlement program in 2017, concerned about his inability to work and his family's future financial situation. As part of his application, he saw a doctor and underwent a series of tests aimed at assessing his cognitive function, including his language, learning, and memory. The doctor concluded that Kevin had experienced cognitive decline consistent with mild dementia. But the NFL rejected Kevin's compensation claim, arguing that his doctor hadn't adjusted his cognitive function test scores for his race in a practice known as "race norming."

Race norming is a practice in which cognitive test scores are adjusted for Black people relative to White people, founded on the assumption that Black people start with lower cognitive skills at baseline. The use of race norming means that a Black person needs to fall to a lower level of cognitive function relative to a White person in order to be considered to have experienced the same amount of cognitive decline. After race norming of Kevin's cognitive test score, he no longer qualified for the NFL's compensation scheme.

"I felt so betrayed and I still feel that way. Two different systems? How can that be okay? White people should be upset too. Not just me, not just Black people. Because if the shoe was on the other foot, you wouldn't like it," Kevin told ABC News.

In August 2020, Kevin and another former NFL player, Najeh Davenport, sued the NFL, claiming that it had deliberately manipulated former players' cognitive test scores in a way that reduced the likelihood of Black former players qualifying for compensation under its settlement scheme. The lawsuit was dismissed by a judge in March 2021, who instead ordered the NFL and the lead lawyer in the settlement to resolve the issue through mediation. Three months later, on June 2, the NFL pledged to stop settling head injury lawsuits using race norming and to review past scores for potential racial bias.[181]

That isn't the end of the story, though. On the same day that the NFL announced that it would stop using the controversial practice to settle players' head injury claims, Katherine Possin, a neuropsychologist at the University of California, San Francisco, told me that race norming of cognitive tests was still a fairly common practice among doctors in the US. "I would say it [race norming] is fairly often used," Possin told me. "There's no study that I'm aware of that surveys neuropsychologists [on their practices], but these are well respected norms," she said.

When I interviewed Possin, I was in the midst of reporting an article for *New Scientist*, investigating racism in medicine. During the investigation, and in the months following the article's publication, I learned some shocking things I want to share here. We have already seen how racist behaviors exhibited by individual doctors or other health care workers—interpersonal racism within medicine—can influence patient experiences and health outcomes. Yet, as I discovered during my investigation, this is only a small part of a much larger problem. Racism in Western medicine is *systemic*. Whether or not an individual doctor exhibits racist behavior is almost irrelevant if we consider that all doctors and other health care workers are operating within a wider system of medical education, guidelines, and practices in which racism is deeply entrenched. The use of race norming by the NFL in the US, to which Kevin's story drew public attention, provides a perfect example of this. Race norming of neuropsychological tests wasn't something that the NFL came up with out of nowhere; the practice already existed in medicine.[182]

Specifically, the practice that the NFL was using was established in 1991 and updated in 2004 by Robert Heaton, a clinical psychology researcher at the University of California, San Diego, as a way to try to account for the fact that African American people tended to score lower than White people on cognitive tests. Subsequent research has shown that adjusting cognitive test performance to take social factors such as education quality into account significantly reduces this variance by race. This fact is even acknowledged by Heaton and his colleagues in one of their research articles, in which they note that "racial/ethnicity disparities on neuropsychological testing may well be accounted for by quality of education [. . .], literacy, acculturation, and other background differences as opposed to any direct result of race/ethnicity, *per se*." It is for these exact reasons that Possin, the neuropsychologist we just heard from, believes that race norming is extremely problematic.[183]

"Being Black in America is on average associated with a less privileged social experience," Possin told me, pointing out that Black people in the US are more likely than White people to experience poverty, early life adversity, and discrimination, and to be less comfortable within a cognitive testing environment that she noted "has historically been used to discriminate against Black people." She believes it is these factors that explain average racial differences in cognitive test scores, rather than something inherent about race. As a result, she is one of several neuropsychologists in the US calling for the development of non-race-based approaches to diagnosis. In light of these calls for change within the neuropsychology field and in the wake of the NFL's decision to stop using race norming, I was curious to know where Heaton himself stood on the issue, a few decades after he had first established these so-called "Heaton norms."[184]

Unfortunately, Heaton didn't agree to be interviewed when I reached out to him. However, he did email me a few comments, which he said he had also shared with "others in the media," and he offered to answer any additional questions I had by email. "I really don't have time to enter this fray, which has become far more political than science-based, but would try to answer any questions you may have," he wrote in his response. During our exchange, Heaton acknowledged that observed differences in test performance between subgroups of the US population might be explained by racial discrimination, stressful life experiences, a lack of consistent access to good nutrition and health care, and limited educational opportunities, but he argued that measuring these factors directly would be too difficult. "These are extremely difficult to measure, quantify and 'correct for' in interpreting test results," he said. "The fact remains, that a very substantial amount of variability in the test performance of normal adults can be 'explained' (accounted for) by the demographic variables of age, education, sex and race/ethnicity (together), so our best available norms 'correct' for these characteristics," Heaton wrote to me.

Possin disagrees. "Race is a crude proxy for lifelong social experience," she told me. "Genetic differences in cognition do not follow these race lines. So correcting for lifelong social experience with something very crude, like race, is not precise medicine," she said. It also perpetuates the false idea that Black people are innately less intelligent than White people, she added. "That's a big problem." Not only that but race norming may additionally contribute to racial health disparities by making it more difficult for Black people to get diagnosed with illnesses associated with cognitive decline, such as dementia, as in Kevin's case.

In order for race norming to be eliminated from cognitive testing, Possin told me she thought it would be helpful if prominent organizations, such as the International Neuropsychological Society and the American Psychological Association, took a stance on the issue. But both these organizations declined to do so when I asked them about it in 2021, following the NFL's pledge to stop using race norming. "The INS does not have guidelines nor take a stance [on] race norming," said the INS at the time. "We are a global organization and focus on topics that are applicable around the world." In a similar vein, the APA responded: "The APA has no official position (or 'stance' as you put it) on race-based norming and cognitive testing."

I would learn that receiving these sorts of responses wasn't unusual when I pressed medical and scientific organizations to comment on race adjustments, which are ubiquitous in Western medicine.

My younger sister is an elite 400-meter sprinter who has competed internationally for Great Britain. She and I are very close and even though we now live in different countries, we still talk almost every day. When I was still living in the UK, she was just a train ride away and we used to visit each other often. I think it was probably during one of those visits, in early 2020, that she started telling me about some blood test results she had recently received. Her GP had told her that her creatinine level was a bit higher than normal—a potential indicator of a kidney problem. That wasn't particularly surprising to me or to her; creatinine is a waste product produced by muscles, and so athletes, who tend to be more muscular on average, commonly have higher than average levels of the compound in their blood without this necessarily being associated with kidney problems. She said she had also shared her blood test results with a sports doctor, to get an opinion from someone more experienced with athletes. He confirmed that creatinine is derived from muscle metabolism and that levels are proportional to muscle mass. He also shared with her a list of factors, which he said could be responsible for raised creatinine levels. In addition to muscle mass, diet, and certain health conditions, one of the factors he listed was "Afro-Caribbean race," my sister explained. "Could my race be affecting my creatinine level?" she asked me.

I was about to stumble on an answer to my sister's question. I was at the beginning of my investigation into what I now refer to as "race-based medicine"—the practice of adjusting medical tests based on a person's race or ethnicity. I had first learned about it many years earlier, in a 2015 TED talk by US academic and author Dorothy Roberts, but I had assumed it would have been a thing of the past by this point; in her talk, Roberts rightly points out that race is a social rather than a biological construct, which therefore has no place in twenty-first-century medical tests. I soon discovered, though, that contrary to my belief, race-based medicine was—and still is—alive and well.[185]

My first clue came in the summer of 2021, before I had started looking into race norming or Kevin Henry's experience with the NFL. It was instead in a US-based study I came across that highlighted issues associated with a widespread practice of adjusting routine kidney test results based on a person's race. In brief: if a doctor wants to assess a patient's kidney health, they will usually start with a test that measures the level of the waste product creatinine in their blood—as it turned out, the same blood test that my sister had in the UK the previous year. The doctor or lab technician conducting the analysis will then plug that blood test result into an equation that calculates the patient's estimated glomerular filtration rate (eGFR), which is the estimated rate at which the kidneys filter waste—an important indicator of how well the kidneys are functioning. Broadly speaking, more creatinine in the blood suggests a lower filtration rate by the kidneys (this is a bit of an oversimplification, but it will do for our purposes). If a patient's eGFR is too low, that could therefore be a sign of a kidney problem. In the worst cases, when the kidneys aren't functioning correctly, toxic waste products build up in the body—a condition that is fatal without treatment such as dialysis or a kidney transplant.[186]

What I gleaned from the study was that the most widely used eGFR equations globally, including in the US, included a specific numerical adjustment or multiplier that was applied to increase eGFR values for Black people. After the application of this Black race adjustment by a doctor or lab technician, a Black patient would end up with a higher eGFR compared with a non-Black patient who had received the same exact blood test results. Not only that, but the study showed that the removal of the race adjustment could improve the accuracy of kidney failure risk prediction among Black adults at high risk. Put another way, it indicated that the use of race adjustment might be reducing the accuracy of kidney function tests for Black people, by overestimating their kidney health.

As I went over the details of the study, my conversation with my sister came flooding back to me. I remembered that she'd had a similar test done in the UK the previous year. I called her and asked her about it again. This time, she shared some photographs of her test results with me, and I could see that it was indeed an eGFR test that she'd had. Printed in brackets above her results was the phrase "If Black multiply result by 1.21."

Staring at my sister's test result sheet, I had so many questions. Where did the Black race adjustment in eGFR come from? How many other countries have similar medical guidance? If race is a social construct, why is it being treated as a biological one? Are there similar race adjustments still being used in other areas of medicine? Might this be causing harm to Black patients? What is the medical definition of "Black," anyway? Is it based on a patient's self-identified race or is it based on the doctor's assumptions about the patient's race? What happens if a patient identifies as Mixed?

The list went on. Fortunately, I was by no means the first person to raise any of these questions. In addition to Dorothy Roberts, kidney doctors such as Vanessa Grubbs and Nwamaka Eneanya in the US have long been calling for the elimination of race-based medical practices entirely, highlighting the lack of evidence to support them. Their calls received renewed attention in the summer of 2020, as medical institutions clamored to publish "Black Lives Matter" statements in the wake of George Floyd's murder and worldwide anti-racism protests. Students who were still being taught race-based practices in medical school began to increasingly question their teachers. Among those asking questions was Naomi Nkinsi, a medical student at the University of Washington.[187]

When I interviewed her in the summer of 2021, Nkinsi patiently walked me through the history behind various race adjustments used in medicine and filled me in on the work that she and others had been doing in recent years to try to bring about change. She explained that there were two equations commonly used in medicine to calculate eGFR, both of which included adjustments for Black race. The first of these equations, called MDRD, was developed in the 1990s. "What this equation does, it looks at a variety of factors that can impact someone's kidney functioning. And one of those factors that is built into the equation is actually Black race," Nkinsi told me. "So, if someone is identified as being Black—and there's no specification whether they're self-identified or if the physician says, 'Oh, this person looks Black'—then their eGFR calculation is adjusted by 1.21," she explained. I immediately recognized those three digits from my sister's test result sheet.

The three digits that constitute the Black race adjustment in the MDRD equation originate from a small US study by the same name conducted in 1999, which found that study participants who self-identified as African American had higher levels of creatinine in their blood on average compared to those who identified as White. "From this they said, 'Oh, well if they have a higher creatinine, it must mean that Black people have a higher muscle mass,'" Nkinsi told me. However, the MDRD study included only 1,628 participants, only 197 of whom identified as African American. "It's based on this one observation that they found out of a very, very small population," said Nkinsi. Moreover, there is very limited evidence to support the secondary assumption that Black people have more muscle mass compared to White people, or to people belonging to other racial or ethnic groups for that matter. That assumption is also at complete odds with the scientific understanding of race as socially constructed. "I do not think there is any evidence that race is related to muscle mass. I think there are poor attempts at showing evidence that this is the case," Nkinsi emphasized.[188]

In 2009, an updated eGFR equation—the second of the two most widely used equations—was developed based on a larger study called CKD-EPI. But the assumption that it was necessary to adjust for Black race was carried through from the MDRD study, Nkinsi explained. Indeed, both the MDRD and CKD-EPI equations for calculating eGFR contain a Black race multiplier (in the MDRD equation the multiplier is 1.212, whereas in the CKD-EPI equation it is 1.159). "So now, we have all of these aspects of medicine that are all dependent on these equations that are now being recognized to be built on faulty science," said Nkinsi. Not just faulty, but potentially harmful.[189]

Around the same time I met Nkinsi, I came across preliminary research from the UK led by kidney doctors Rouvick Gama and Kate Bramham at King's College London. Their research showed that eGFR equations with race adjustments overestimated actual GFR in Black patients, compared to results using a more invasive but more accurate method. I reached out to Gama and Bramham and they agreed to meet me over a video call to discuss the implications of their findings. Gama told me that the overestimation of kidney health due to race adjustment could have serious consequences for Black patients. "It could lead to delay in diagnosis of chronic kidney disease," he said, and therefore delays in treatment—something that is reflected in UK health statistics. "If you're of Black ethnicity, you're three- to five-fold more likely to end up with end-stage kidney disease," Bramham told me. "Almost certainly we're not recognizing it enough," she said.

There is a similar picture in the US, where, according to the National Kidney Foundation, Black or African American people are more than three times as likely as White people to develop kidney failure. Nkinsi told me she thought that the race adjustment was a contributing factor to that disparity. "What it means is that we're missing kidney disease in Black people," she said. "In order for a Black person to be identified as having kidneys that are sick, they actually need to be sicker than a non-Black person," she emphasized. "That means that you have later access to specialized care, later access to things like transplant, later access to things like Medicaid coverage for kidney care. That leads to worse health outcomes for Black people," she said.

In the US, it is estimated that thirteen people die every day while waiting for a kidney transplant. Black or African American transplant candidates wait longer on average than White transplant candidates for kidney and other organ transplants—and use of race adjustment in calculating eGFR has meant that their chances of getting onto the transplant list in the first place have been limited. I didn't know it at the time, but as Nkinsi and I were talking, Jordan Crowley, a college student in New York who was born with just one partially functioning kidney, was having to wait to get onto the transplant list for that very reason. Jordan identifies as Mixed; he has one Black grandparent and one White grandparent on each side of his family. When assessing his kidney health, his medical providers racialized him as Black and applied the notorious race adjustment to his eGFR calculation. As a result, he didn't qualify for the transplant list. Yet if his doctors had racialized him as White, he would have.

Jennifer Tsai, a US-based physician and writer, drew attention to Jordan's case in an article published in *Slate* a few weeks after my first conversation with Nkinsi. Tsai was introduced to Jordan by his uncle, Jay Kaufman—the epidemiologist at McGill University. Tsai and Kaufman had been collaborating on a study, which has since been published in a scientific journal. It estimated that removal of the race adjustment from eGFR calculations in the US would have resulted in 31,000 additional Black patients becoming eligible for a transplant from 2015 to 2018.[190]

On October 1, 2021, Jordan filed a lawsuit against his medical providers alleging racial discrimination. His ordeal encapsulates the absurdity of using race to determine biology and illustrates how devastating the consequences can be to a person's life. The case filing sums it up: "It has been known for decades in the scientific and medical community that the use of this medical algorithm is not grounded in science, and that race is an arbitrary and inaccurate indicator of a person's kidney health." *Why on Earth, then, was the algorithm still being used?* That was the question I wanted to get to the bottom of. The answer seemed to lie in the world of medical guidelines.[191]

As I learned more about the use of race adjustment in eGFR and its potential harms for Black people, I began reaching out to some of the health bodies who I thought might be responsible for setting this sort of medical guidance. That's when I began to realize the scale of the problem, and why seemingly little progress had been made in this area despite so many medical students, doctors, and academics calling for change.

Toward the end of May 2021, I sent a media request to the CDC in the US, mainly to clarify what their current guidance was for the calculation of eGFR in adults and to ask if they recommended the use of race adjustment. I received a friendly response explaining that the CDC didn't issue guidelines on eGFR calculation, but that the current guidelines in the US came from Kidney Disease Improving Global Outcomes (KDIGO)—"a global organization that develops and implements evidence-based clinical practice guidelines in kidney disease," according to my contact at the CDC, who also kindly sent me a link to KDIGO's website. I clicked on the link and downloaded the most up-to-date version of KDIGO's guideline for the evaluation and management of chronic kidney disease. The guideline was laden with medical jargon and didn't make for the most thrilling of evening reads, but I soon landed on the section I was searching for. I was once again greeted with the same instruction that had been printed on my sister's eGFR results, to multiply the patient's eGFR by a specific numerical factor "if Black."[1192]

My heart sank. I knew this was a global organization, so its guidance would be relied upon internationally. I drafted an email to send to KDIGO's communications team. I asked if they could confirm that the guideline I had been looking at was their most up-to-date one, and to explain what the scientific rationale was behind their recommendation to use race adjustment in eGFR calculations. I received a prompt response, confirming that the guideline was up to date and curtly stating: "KDIGO is not in a position to comment on the rationale used to determine the adjustment in eGFR calculations." I felt a bit deflated, but I didn't give up.

When I had spoken to Nkinsi, she had told me about how she had questioned her teachers at medical school the first time her class was taught about eGFR. "When we were presented in lecture with, 'Okay, so we're using the MDRD equation, you have to adjust for race,' I was like, 'Wait a minute, that doesn't make sense,'" Nkinsi told me. "In our physiology lecture, no one ever said that Black kidneys work different," she pointed out. "What is the physiology?" No one could give Nkinsi a straight answer. "They were just kind of like, 'Well, this is how we do things,'" she said. Nkinsi wasn't satisfied, and started digging, which is how she discovered the concerning origin of the race adjustment in the MDRD study. "One study can change the trajectory of so much of what we do, because it's further grandfathered in," she sighed.

"Medicine and scientists, we are supposed to pride ourselves in gaining new information, looking at how the world is changing, getting new data and saying, 'Okay, what are we doing?' Nothing we do is solid, right? Everything is supposed to be able to be questioned, right? But, unfortunately, medicine is very hierarchical. It's very much based on this kind of false meritocracy, where for the longest time, White men are running everything, you can't question your superiors, medical students and younger physicians are supposed to just take things as they're taught. You're taught these equations, you're taught these facts, and you just kind of perpetuate them," said Nkinsi. "You're not really taught to question where this information is coming from."

Thankfully, Nkinsi *did* question things, and when she realized they were fishy, she started to demand action from UW Medicine, the medical school where she was enrolled at the University of Washington. In response to Nkinsi's calls, in 2020 UW Medicine announced it would transition away from the use of race adjustment in eGFR calculations. They were not the first or the last medical institution in the US to make this move. Other institutions, such as Beth Israel Deaconess Medical Center in Massachusetts, Massachusetts General Brigham, Mount Sinai Health System in New York, Zuckerberg San Francisco General Hospital, and Vanderbilt University Medical Center in Tennessee also abolished the race adjustment, and in 2020 the National Kidney Foundation and the American Society of Nephrology established a joint task force to "examine the inclusion of race in the estimation of GFR."[193]

Inspired by Nkinsi's success, I continued reaching out to various health bodies in other countries regarding their stances on race adjustment. I received a response from the UK's National Institute for Health and Care Excellence (NICE) in June 2021, regarding its guideline on calculating eGFR, which at the time recommended applying "a correction factor to GFR values [. . .] for people of African-Caribbean or African family origin." NICE told me it was in the process of updating that guideline and reassured me that "there was some consideration in the update about adjustment based on the characteristics you have highlighted." In other words, NICE was considering updating its recommendation to adjust eGFR for race. My contact at NICE's press office shared a link to the draft of the updated guidance, which was due to be published two months later, in August 2021. When I read through it, I was disappointed to see that the recommendation to "correct" eGFR for people of African-Caribbean or African family origin was still there. It was accompanied by a note saying that future research should explore the use of "factors other than ethnicity" as biological markers. At this point, I decided that the best thing I could do was put my head down and write up my report for *New Scientist*.[194]

A few weeks after my article was published, Gama and Bramham—the UK-based nephrologists—got in contact to let me know that their preliminary study, highlighting the potential harms of race adjustment in eGFR to Black patients in the UK, had been published in a scientific journal. It was August 2021, and I knew that NICE was due to publish its updated guideline soon. I forwarded Gama and Bramham's research article to my contact at the NICE press office and asked whether NICE was aware of their findings and if it planned to change its guideline on the use of race adjustment in eGFR. After a couple of days, the response I had been waiting for finally came. NICE had made the decision to remove the race adjustment from its recommendations on calculating eGFR. It would publish its updated guideline the following morning. I was delighted, as were Gama and Bramham. "This is a really important opportunity to stop race-based medicine," Bramham told me at the time.[195]

Indeed, it finally felt as if there was movement in the right direction. I wrote up a quick report on the guideline change for *New Scientist*, reaching out to the UK Kidney Association to get their reaction to the news. "We welcome and support this change," said Paul Cockwell, president of the association. "Ethnicity and race are social constructs and do not match genetic categories," he said. "Adjusting for kidney function based on ethnicity could lead to an overestimation of kidney function and potential inequality in delivery of care," Cockwell concluded.[196]

On the other side of the Atlantic, there had also been some movement on this issue. A few weeks after NICE published its updated guideline, the National Kidney Foundation and the American Society for Nephrology formally established a consensus against the use of race adjustment in kidney function equations in the US. Nkinsi and I caught up afterward. "It is exciting to see that such a big change is being made," she told me. "But the concern is that institutions will see this as, 'Okay, now we're not racist anymore,'" she added. "There's a concern that once institutions around the country are making this change, they're not going to go deeper and look at, what are other ways that Black patients are getting inadequate kidney care?" she explained.[197]

The change is unlikely to be instantaneous either. Kaufman, the epidemiologist at McGill, explained to me that while the National Kidney Foundation and American Society for Nephrology are influential organizations in the US, they don't have any power to dictate what doctors do. "It's not like a law," he said. "This is just a professional recommendation." That means it is up to individual medical institutions to adopt the change. "I'm expecting they're going to drag their feet," Kaufman concluded.

In the UK, it will probably also take time for medical practice across the country to catch up with the updated guideline from NICE. In January 2023, more than a year after it updated its guideline on eGFR to remove the race adjustment, my sister forwarded me her results from a recent medical checkup. Under her eGFR result was the now all too familiar phrase "If patient is black, multiply by 1.21." Kate Bramham explained to me at the time that—at least in England where my sister is based—laboratories "are often not changing their practices until they upgrade their software." As of February 2023, most labs in England had changed their practices, according to Bramham, so perhaps my sister was unlucky. Fortunately for my sister, her creatinine level and eGFR were considered normal given that she is a muscular athlete, irrespective of race. For others, though—such as Jordan Crowley in the US—the use of race adjustment literally means the difference between being eligible for a kidney transplant or not.

Although progress might be slow, the guideline changes in the US and UK were still meaningful and they have been followed by further positive developments. As an example: in June 2022 the board of the Organ Procurement and Transplantation Network, which connects transplant centers and develops policies on transplantation in the US, approved a measure requiring transplant hospitals to use a "race-neutral calculation" when estimating a patient's kidney function. This measure would seem to be beneficial for Jordan's legal case against his medical providers, which is ongoing at the time of writing.[198]

"This is just the beginning," Nkinsi told me. She also pointed out that race adjustment in eGFR represented just one of many persistent race-based medical practices—a growing number of which have come under criticism both within and beyond the medical community in recent years.

Thankfully, different fields of medicine are one by one joining calls to eliminate or at least reconsider the use of race-based medical practices—some of which have unequivocally racist origins. For example, widely used race adjustments in spirometry tests for lung function originate from a suggestion by US doctor and slaveholder Samuel Cartwright back in the 1850s. He claimed that Black people had lower lung capacity than White people and were therefore only healthy when enslaved (you may recognize Cartwright from our brief discussion of "drapetomania," a disease he invented to pathologize enslaved Black people who tried to escape). To reiterate: a racist notion promulgated by a nineteenth-century apologist for slavery is still affecting people's health care today. But lung doctors, such as Alexander Moffett at the University of Pennsylvania, are working to change that.[199]

When I interviewed Moffett in 2021, he had recently presented findings from a preliminary study—on which he collaborated with kidney doctor Nwamaka Eneanya among other colleagues—at the 2021 American Thoracic Society International Conference. The study analyzed data from more than 14,000 lung function tests in the US and showed that removing race adjustments from the interpretation of the tests saw the number of people correctly diagnosed with a lung defect increase from about 60 to 82 percent. These results suggest that adjusting for race in lung function tests may underestimate the severity of lung disease among Black patients, Moffett explained to me. "We're assuming that their lung function should be worse and using that as a way to approach the diagnosis," he said. That assumption is flawed. "Everything we've learned about race in the last fifty years has invalidated this," said Moffett. He and his colleagues are working on developing a method for interpreting lung function test results, which isn't built upon unfounded assumptions about racial difference.[200]

At the time, joint European Respiratory Society (ERS) and American Thoracic Society (ATS) technical standards still recommended the use of different lung function test equations depending on a person's ethnicity. While I was working on my article, I asked the ERS about the scientific rationale behind this recommendation. Here's part of the long response I received, which the ERS attributed to "authors of the guidelines paper": "Traditionally population-specific reference equations have been developed to account for observed differences between people living in different geographical areas. However, the reasons for observed differences in lung function between people around the world are multifactorial and not fully understood."[201]

I also reached out to the ATS at the time, who were more forthright in acknowledging the problems with the recommendation. "Ingrained in lung function interpretation is the long-standing assumption that the observed differences across racial and ethnic populations are biologically based," the ATS said in a statement. "There is increasing recognition that race and ethnicity are sociopolitical constructs which are more reflective of the differing social and environmental conditions across populations than representative of true biological differences," it noted, adding that it was "committed to leading action to address racism in medicine and eliminate the misuse of race and ethnicity in clinical decision making" and that it had "convened a workshop to critically evaluate current guidelines." In 2023, the ATS issued an official statement for clinicians explaining why race and ethnicity should no longer be considered factors in interpreting the results of spirometry. The statement was endorsed by the ERS.[202]

Some medical fields have moved more quickly than others when it comes to reviewing and updating race-based practices. As we discovered earlier, the VBAC calculator—a tool commonly used by doctors in the US for assessing the safety of a vaginal birth after a previous cesarean section—used to contain adjustments for African American and Hispanic women, which reduced their predicted chances of a successful VBAC. But that changed in May 2021, after William Grobman at Northwestern University in Illinois and colleagues—the researchers who originally developed the calculator—published an updated version with the race and ethnicity adjustments removed. Grobman and colleagues cited equity concerns raised in an earlier paper from 2019 as the reason they made the change. The lead author of that 2019 paper was Darshali Vyas, who was then a medical student and is now a resident physician at Massachusetts General Hospital. Reading Vyas's paper was how I first learned about the concerning origins of the unevidenced claim made by D'lissa Parkes's doctor in the UK, that Black women's pelvises are shaped differently to White women's.

I caught up with Vyas at the end of 2021 and we reflected on the sometimes fast, but more often painfully slow, changes that were taking place across different fields of medicine, starting with obstetrics and gynecology and the recent changes that had been made to the VBAC calculator. "I've gotten to talk to Dr. Grobman and the group and I feel very optimistic about the changes they made," Vyas told me. "I also think it's super powerful that it was the same group that created the first calculator, who then took a critique, really thought about it, and revalidated the tool without it and showed that it was just as robust and just as useful a tool," she added. "I think that is meaningful. And it's honestly what good equity work, I think, should look like." Vyas explained that unlike with the calculation of eGFR, which depends on individual medical institutions taking the initiative to remove the use of race adjustment, the VBAC calculator is accessed via a single online source and so was much easier and quicker to change. "They've updated the official calculator online, so it is now fully replaced," she said.

Like Nkinsi, Vyas was in medical school when she first became concerned about the use of race-based algorithms in clinical practice. She noticed a contradiction between how she was being taught about race academically—as a social construct with no biological basis—and how race was being used by the doctors she was learning from on the wards. She joined a group of students at her university called the Harvard Medical School Racial Justice Coalition, which is where she realized that a lot of her classmates shared her concerns. "I think a lot of my classmates felt a similar tension," she told me. The following year, Vyas started a core clinical rotation in obstetrics and gynecology, which is where she first encountered the VBAC calculator. "Because I had this framework from social justice organizing and from conversations with my classmates before I got to the wards, I think I was very primed to pick up on this example of race correction," she reflected.

As Vyas transitioned from being a medical student into being a doctor, she came across more and more examples of race corrections or adjustments being used across different fields of medicine. In addition to various medical fields having individual moments of reckoning on the specific race-based medical practices within their respective domains, she wanted to stimulate a much broader discussion about the problem with the way race is used within medicine overall. "I think we need some sort of guiding principles in medicine, overall, about how we should and should not think about race," said Vyas.

Indeed, it's clear from what happened to D'lissa Parkes that inaccurate notions about race in medicine can cause harm without even being embedded into algorithms. US-based physician Richard Garcia alluded to this in a 2004 commentary published in the medical journal *Pediatrics*, in which he wrote about his childhood friend Lela, who wasn't diagnosed with cystic fibrosis until she was eight years old. "Had she been a white child, or had no visible 'race' at all, she probably would have gotten the correct diagnosis and treatment much earlier. Only when she was 8 did a radiologist, who had never seen her face to face, notice her chest radiograph and ask, 'Who's the kid with CF [cystic fibrosis]?'" Garcia wrote.[203]

According to the Cystic Fibrosis Foundation, a US-based nonprofit organization, about 5 percent of people with cystic fibrosis in the US identify as Black. Yet the condition is often inaccurately perceived as an exclusively "White disease," which may contribute to underdiagnosis and poorer outcomes with the condition among people of color. Sickle cell disease (the group of blood disorders that includes sickle cell anemia and other related conditions) has been similarly racialized as a "Black disease," which in addition to being inaccurate has had the added disadvantage of contributing to the marginalization and dismissal of sickle cell patients due to racism. Sickle cell is known to be associated with extremely painful episodes called crises, yet in both the US and UK there is evidence that Black patients requesting pain relief are often disbelieved by health care workers and stereotyped as drug-seeking. "I definitely feel that race does play a significant role in how patients are treated, especially in A&E. I think there is the misconception that the drug-seeking patients are back here again," June Okochi, a sickle cell patient representative in the UK, told a 2021 inquiry into avoidable deaths and failures of care for sickle cell patients. On top of the health harms of receiving inadequate care, the stress associated with this racist treatment results in many patients avoiding health care altogether, further contributing to health inequity.[204]

Historically, the sickle cell gene occurred more often in parts of the world where malaria was more common, including equatorial Africa, the Middle East, and India, because of its protective effect against malaria. While that is helpful for understanding population genetics, it doesn't provide an accurate way of diagnosing an individual patient. Like sickle cell, cystic fibrosis is also a genetic condition and, nowadays, the most accurate way to identify genetic conditions like these is to leave the racialized assumptions at the door and simply test for the presence of mutations in the relevant genes.[205]

In general, vague notions of race don't provide accurate ways of diagnosing people, even in cases where they may have been legitimized through their incorporation into medical algorithms. In these cases, race is often being used as a proxy for something else, such as muscle mass in the case of the eGFR equations. Nkinsi summed up the problem: "We're using race as a proxy for other things, instead of measuring those things directly."

After my article about the problems with race-based medicine was published, a few readers wrote to *New Scientist* with critical comments and questions. The question I think most of them were trying to get at was: how *should* doctors consider race? Surely, they couldn't be expected to ignore race entirely, since it is clearly associated with people's health outcomes. I definitely don't think the movement away from race-based medicine should be a movement toward color-blind medicine. But I do think there is a big difference between using race as a proxy for health and acknowledging racism as an underlying process contributing to health inequity.

When I put these points to her, Vyas argued that "I would actually say that putting race into tools is a form of race-blind medicine, because you're hiding disparities." She went on, "You're sort of shaping your tools around existing disparities, rather than actually addressing them at their root cause," and further, "I firmly believe that we should always continue to study race and its effects on health outcomes. We should study and understand how racism affects health outcomes. And we should really be quite explicit in understanding how racism is impacting our patients." In the end, "Where we draw a distinction is between studying racial inequities and building our tools in a way that could just exacerbate them or hide those disparities completely," she concluded.

Nkinsi echoed these comments. "We want people to practice antiracist medicine, not color-blind medicine," she told me. "What that means is that you recognize that race isn't biological, it's a social construct. And that these are very real social dynamics in how people are treated and how they experience life," she said. As a physician that means thinking about *racism* and how that affects your patients, rather than categorizing them based on pseudoscientific assumptions about race.

Racism has a profound effect both on people's health and on the way they are treated within the health care system when they become ill. The persistence of race-based medicine in the form of algorithms is testament to just how deeply racism is entrenched in Western medical practice. In August 2020, just over a year after her paper about the VBAC calculator was published, Vyas coauthored another paper in the *New England Journal of Medicine*, which detailed examples of race-based algorithms from various medical fields and highlighted their harms. "By embedding race into the basic data and decisions of health care, these algorithms propagate race-based medicine," Vyas and colleagues argued. Their paper, titled "Hidden in Plain Sight—Reconsidering the Use of Race Correction in Clinical Algorithms," played a key role in helping to catapult the issue of race-based medicine into mainstream medical and scientific discussions. It has since received more than six hundred citations, including from US senators Elizabeth Warren, Ron Wyden, and Cory Booker, and Congresswoman Barbara Lee, who a month after its publication requested a national review of the use of race-based medical algorithms. "In order to reduce health disparities among communities of color, we must ensure that medicine and public health organizations take a staunchly anti-racist approach to medical care and reevaluate the ways in which current practices, including the use of race-based algorithms, could be worsening outcomes for people of color," the four politicians wrote in a letter to the Agency for Healthcare Research and Quality in September 2020.[206]

"I think many people would accurately say this is overdue and certainly it's important to highlight there have been scholars, Black scholars, Black doctors, Black activists, who have been sounding the call around race-based medicine for a long time," Vyas told me. "It might be something about the past few years—a global pandemic that has really highlighted these disparities, as well as mounting police violence, and police killings of Black people—that maybe has cast this in a new light that has gained more momentum, but certainly these conversations preceded me," she acknowledged.

Vyas's 2020 paper lists many more examples of race-based algorithms than I have space to write about—and during the process of writing this book I have continued to discover more. After my *New Scientist* investigation was published, I received a message via social media from Dipesh Gopal, a GP at Queen Mary University of London. He told me he had been pleased to see my article about ethnicity-based kidney test adjustment being removed from UK medical guidance issued by NICE, and he mentioned that he had been trying to encourage NICE to review another one of its guidelines with less success. We met over a call so that he could elaborate.

He explained that the NICE guideline on the treatment of hypertension or high blood pressure recommends different treatments for people depending on their ethnicity. As of 2022, the guideline suggests that doctors should prescribe drugs called ACE inhibitors to people under the age of fifty-five with high blood pressure—unless they are of "black African or African-Caribbean family origin," in which case it recommends prescribing different drugs. Gopal told me he and his colleagues had written to NICE twice requesting that the guidance be urgently reviewed. NICE declined in both cases, responding that evidence indicated there were "clinically meaningful differences in the effectiveness of treatments for individuals in these family origin subgroups." However, Gopal and other doctors and academics—including epidemiologist Jay Kaufman, and physician and writer Jennifer Tsai—dispute this evidence.[207]

Kaufman pointed out several flaws in this evidence in an email sent to Gopal and others following NICE's decision not to review its guidance. He noted that while it was true that some studies had found that Black people on average have less activity of an enzyme called renin in their blood—an indicator that they may respond less well to treatment with ACE inhibitors—those studies also found a large amount of overlap in renin activity levels between people from different racial and ethnic groups. Based on data from one such study, if you select any Black or White patient at random, there is a one-in-three chance that the Black patient will have a higher plasma renin activity than the White patient, said Kaufman. This seemed to run counter to another argument made by NICE in its response to Gopal and colleagues, that ethnicity provided "a pragmatic means of estimating a patient's renin status."[208]

When we spoke, Gopal also noted that it isn't clear from the guideline how doctors are expected to determine a patient's ethnicity in practice, or what to do in cases where people have more than one racial or ethnic identity. Tsai and I decided to write an article to highlight this issue—and to call for NICE and other medical institutions globally to systematically review race-based recommendations across their guidelines once and for all. In response to Gopal and his colleagues, and to the content of our article, NICE acknowledged that "there is not a clear-cut biological and genetic homogeneity amongst all Black and White people" and that "the guideline does not account for people with mixed heritage." But it said performing the relevant tests on everyone wasn't possible due to "the expense, and the additional time." Our article was published in *New Scientist* in November 2021 and, unfortunately, remains relevant.[209]

A few weeks before our article was published, in October 2021, the House Ways and Means Committee—a US congressional committee—released a report probing what it described as the "misuse of race in clinical algorithms," citing Vyas's 2020 paper among others. The report noted that although there is desire within the medical community to improve current practices, "divergent opinions on the appropriate use of race and ethnicity in CDSTs [clinical decision support tools] make it difficult to develop broad consensus that would uniformly transform behavior across the medical community."[210]

I am slightly more optimistic. In the process of writing this book, I have gotten to know so many inspiring patient advocates, medical students, doctors, academics, and other activists who have dedicated themselves to advancing health and medical equity, and I am excited and passionate about the vision of anti-racist medicine that they and so many others are working toward. I understand that racism can be very sticky and therefore slow, painful, and difficult to excise, including from the institution of Western medicine. But I believe it is possible, and we have to start somewhere. As Nkinsi said, this is just the beginning.

In addition to medicine, racism in science also contributes to health inequities. In particular, biases and data gaps in medical research may be entrenching existing racial and ethnic health gaps. And that begins with what is missing—or, rather, who is missing—from health data and medical studies.

PART IV

Data Gaps

CHAPTER 8

The Missing Data

After almost eight years living in the UK, in 2020 I relocated to Berlin, Germany, with my partner. We are thinking about having children and, naturally, I have been doing some research on the subject. Having become aware of the vast racial and ethnic disparities in pregnancy outcomes that exist in the UK and elsewhere, I was curious to learn more about how the situation in Germany compares. Unfortunately, health data disaggregated by race and ethnicity are sparse in Germany. But this missing information doesn't give me any comfort. Far from it: you cannot fix a problem you don't know is there.

Germany is by no means exceptional in this respect. In fact, it is countries such as the UK, the US, and Canada, where the collection of race and ethnicity data is more commonplace, that are the outliers. More than half of the countries in the OECD, an intergovernmental organization comprising mainly wealthy nations, don't routinely collect data on the race and ethnicity of their populations in the context of health—or in any other context for that matter. Among those OECD countries that are members of the European Union, that figure rises to more than 60 percent. Germany doesn't even know how many Black people live in the country, let alone what their pregnancy outcomes are. The situation is similar in other EU member states such as France, where the collection of data on race or ethnicity has been restricted by law since 1978. Living in Berlin, it is easy to see how Europe's recent history presents plenty of cause for caution when it comes to gathering this type of sensitive information (more on this later). But not collecting it at all may be doing more harm than good.[211]

Despite the limited available data, there are still indicators that racial and ethnic health disparities around pregnancy and childbirth are present in EU countries. In the Netherlands, where I was born, a study conducted from 2004 to 2006 found that "non-Western immigrants" had a 30 percent higher risk of developing severe maternal morbidity compared to "Western women." Limited data from France between 2001 and 2006 suggest maternal mortality was higher among non-nationals, with separate data from the five years up to 2001 indicating mortality risk was highest among women from sub-Saharan Africa. In Germany, data collected from three Berlin-based maternity clinics between 2011 and 2012 indicated that outcomes for most perinatal parameters were largely similar for "Turkish immigrant women" and "non-immigrant women," but the data were not disaggregated further. And in any case, separating people into immigrants and non-immigrants based on the country in which they were born is a poor proxy for race or ethnicity in today's world—many people belonging to minoritized racial or ethnic groups are not first-generation immigrants (or even second-generation for that matter). I, for example, identify as Black and Mixed, but I was born and grew up in the Netherlands. Nationality is an equally unhelpful proxy for race or ethnicity for the same reason—I am both Black and British, for example, and being British doesn't somehow erase my Blackness.[212]

My frustration at researching racial and ethnic health inequalities for this book and discovering that the relevant information was often incomplete or completely absent is a big part of what inspired me to dig deeper. I am extremely grateful for the information I was able to access, as this book wouldn't exist without it. But to ignore the missing data would be unscientific. Racism is a global health issue, yet searching the English language academic literature on the subject yields research mainly from a single country: the US. A systematic review led by Yin

Paradies at Deakin University in Australia, which analyzed hundreds of studies on racism and health published between 1983 and 2013, found that the vast majority—about 81 percent—had been conducted in the US. The next largest proportion of research coming from a single country was tied between the UK and Australia, which each accounted for about 3 percent of the studies analyzed, followed by Canada, which accounted for 2 percent. You may also have noticed that many of the examples and studies I have cited in this book so far have come from the US and UK. That is partly due to my own bias as an English-speaking author working with publishing houses based in those two countries. But it is also connected to the relative abundance of data in those places—something I suspect may be due to their particular histories of racism and their relatively diverse present-day populations—as well as the relative lack of data in other places.[213]

There is a growing awareness both within the EU and worldwide that the failure to properly disaggregate data based on race and ethnicity causes harm, not only by erasing people's lived experiences, but also by limiting opportunities to identify and address inequities. In September 2020, in the aftermath of worldwide anti-racism protests, the European Commission released a five-year Anti-racism Action Plan for the EU highlighting the need for better data collection by member states, including data disaggregated by "ethnic or racial origin." The following year, in June 2021, a UN report by the Office of the United Nations High Commissioner for Human Rights (OHCHR) called for countries globally who weren't already doing so to start collecting and making publicly available "comprehensive data disaggregated by race or ethnic origin."[214]

Ojeaku Nwabuzo, a senior research officer at the European Network Against Racism (ENAR) based in Brussels, was among the contributors to the UN report. Like me, Nwabuzo is Black and British. Before moving to Belgium to work at the ENAR, a nongovernmental organization working to end structural racism in the EU, she lived in the UK and worked as a policy and research analyst at the Runnymede Trust, a race equality think tank. When she arrived in Belgium, she was struck by the scarcity of what she and her colleagues refer to as "equality data collection." As was the case for me when I moved to Germany, the lack of race and ethnicity data collection in Belgium made her feel invisible, like her existence and experiences in the country weren't fully acknowledged. "You're similar to me," Nwabuzo told me. "You lose your identity when you cross the border and it's bizarre," she continued. "My Black Britishness is completely gone," she said, and I could instantly relate. This isn't only a matter of identity though. Nwabuzo and other equality data experts I spoke to explained to me how detangling data to unveil patterns that may be present among subgroups of a population can provide a powerful way of pinpointing and tackling health and other inequities. As *Invisible Women* author Caroline Criado Perez summed it up in an interview with *Forbes*: "Until you have the data, you can't address the discrimination."[215]

To take an example: when the world was first hit by Covid-19, breaking down case data based on commonly recorded demographic variables, such as age, gender, and occupation, was hugely informative for our understanding of the epidemiology of the disease. Recognizing that the risk of severe illness increased with age, for instance, helped inform government interventions across the globe, such as decisions to prioritize older people for vaccinations. Similarly, countries in which information regarding the race and ethnicity of Covid-19 cases was made public during the early stages of the pandemic, including the US and UK, were among the first to draw attention to concerning trends of people from racial and ethnic minority groups experiencing greater-than-average risk from the disease. And the publication of these kind of data can help drive governments to act; the UK government launched an inquiry into the issue. Later in the pandemic, when the UK started its Covid-19 vaccine rollout, I witnessed firsthand another example of how ethnicity-disaggregated data could be beneficial.

Around the same time that the first people began receiving their Covid-19 shots in the UK, in late 2020 going on to early 2021, a number of surveys suggested that vaccine hesitancy was more common among ethnic minority groups. As the UK's vaccine rollout picked up pace, I wondered whether those initial reports about increased hesitancy were being borne out in the form of ethnic disparities in vaccination rates. But when I searched for data on the ethnicity of people who had so far received the shot, I couldn't find any. As I dug deeper for this information, I started to suspect that the UK government wasn't collecting it. I spoke with several public health leaders, who were critical about the missing data and told me that the information was vital for monitoring and tackling ethnic inequalities in Covid-19 vaccination coverage. By this point, in late January 2021, a preliminary study led by Ben Goldacre at the University of Oxford had indicated that there were indeed ethnic disparities in vaccination in England, and several journalists—myself included—had reported on the lack of official ethnicity-disaggregated information on the Covid-19 vaccine rollout. Soon afterward, NHS England started publishing weekly data on Covid-19 vaccinations by ethnicity. As more and more data emerged from various official sources, patterns of ethnic inequality in vaccination coverage became clearer.[216]

On March 29, 2021, the UK's Office for National Statistics published an analysis indicating lower-than-average vaccination rates among people from Black African, Black Caribbean, Bangladeshi, and Pakistani ethnic groups, with the lowest rates among people from Black ethnic groups.* These and other data were crucial in informing targeted efforts to encourage more people from those groups to get vaccinated. For instance, the day after the Office for National Statistics publication, British actor and comedian Lenny Henry published an open letter and appeared in a public health campaign video along with other Black celebrities appealing to Black Britons to "take the jab." The letter and

* These statistics refer to people aged seventy and above. At the time these data were collected, most of the people who had been offered a Covid-19 vaccine in England were in this age group (people were prioritized for Covid-19 vaccination according to age, from oldest to youngest).

video, part of a broader, government-backed vaccine confidence campaign, acknowledged the legitimate worries and concerns that many Black people have about trusting institutions and authorities. Following this and other targeted campaigns, vaccine confidence increased across all ethnic groups. Although a lot more needs to be done to fully address ethnic health inequalities in the UK (some ethnic disparities in Covid-19 vaccination rates persist at the time of writing), the availability of ethnicity-disaggregated data has undoubtedly been helpful in beginning to tackle the problem. "When you have the data, it's really much harder to ignore," Nwabuzo told me. "Then the policymakers have to do something—or be seen to do something—which helps," she said.[217]

Despite the potential benefits, in many countries—especially in the EU—the collection of race and ethnicity data remains an extremely sensitive topic. In Germany, for instance, the Nazis' use of population registers in organizing the Holocaust has understandably made people especially mindful of the potential dangers of personal data falling into the wrong hands, particularly in medical contexts. "The history in Germany is very clear," Nwabuzo told me. "In France, it's more about their specific ideology on the identity of the country that prevents them, in theory, from collecting information on race. Because they want to see people as French people and not as different ethnic groups," she said. "So, it's a combination of European history and political ideology," Nwabuzo explained.

Concerns about data protection and fostering inclusion are perfectly valid. "There are examples of current and past misuse of information and data collection," Nwabuzo acknowledged, adding that she could understand why some people belonging to racialized groups might be particularly hesitant about handing over personal data to authorities. That said, racism clearly exists in Europe—almost 60 percent of Europeans believe racism is widespread in their country—and continuing to avoid the subject entirely won't help to solve the problem, Nwabuzo pointed out. Instead, she proposed, countries should focus on ensuring that race and ethnicity data are collected in an ethical way, with measures taken to protect people's privacy. This could help to rebuild public trust and enable racial and ethnic health disparities to be detected and dealt with. Her position is in line with that of the EU.

"The EU as a regional, political institution has been very clear that member states should do this kind of data collection," Nwabuzo said. The European Commission has issued several guidelines outlining how member states should collect these data, she told me. Anti-racism organizations, including the ENAR, where Nwabuzo works, have also published guidance on equality data collection in compliance with data protection and privacy laws. "We produced, years ago, guidelines on how to do this data collection very, very simply," she explained. "And, yet, member states don't do it," she sighed. "There's a political unwillingness, because it's very easy to say, 'Oh, we don't need to collect the data, we're doing fine,'" Nwabuzo said. She thinks the EU needs to go further than just presenting guidance and support to member states. "I think there comes a point where they have to really hold them to account," she concluded.[218]

Daris José Lewis Recio, a legal and policy officer at the European Network of Equality Bodies (Equinet), also based in Brussels, agrees. Lewis Recio and his colleagues at Equinet have also been working to improve equality data collection in the EU, and he agreed to speak with me about the issue in a personal capacity (his views don't necessarily reflect those of Equinet). "I think that this is now time to be a bit more strict," he said. The EU could do this by enacting or enabling an infringement procedure for member states that fail to gather data, he suggested, although this wouldn't be straightforward; equality data collection is encouraged but not required under EU law. "I don't think it will reach the point at the EU level at which, for example, they will require member states to collect data on racial and ethnic origin. I don't think so, at least not in the short term," Lewis Recio told me.

There are signs of progress in some countries, though. In November 2021, results from the first large-scale online survey of Black, African, and Afro-diasporic people in Germany were published in an effort funded by the government and led by the civil society organizations Each One Teach One and Citizens for Europe. Daniel Gyamerah, a racism researcher at Each One Teach One in Berlin who is part of the core team behind Afrozensus, spoke about the motivation behind the survey in a YouTube video released shortly before it was published. "Why are we doing this?" asked Gyamerah. "First and foremost, for our communities, for ourselves," he said.[219]

The fact that Afrozensus was community-led was a key strength of the survey, said Lewis Recio, when we spoke. "The communities and the data subjects, you know, the groups concerned, they need to be involved in the process. That's super important," he told me. "They should be able to not only be the target, but also the experts leading [research]," he emphasized. Almost six thousand of the more than one million Black, African, and Afro-diasporic people estimated to be living in Germany participated in the poll. The results highlighted widespread experiences of racism, including in the context of health care. Nwabuzo also praised the research effort and said she hopes that the results from Afrozensus will inspire more systematic data collection and disaggregation by authorities in Germany and in other EU countries.

Even in countries such as the US and UK, where official collection of equality data is more common, there are limitations to the way these data are gathered. In the UK and US, forms asking for race and/or ethnicity information usually provide a set list of options from which people are asked to select the one that best describes them. When I was living in the UK, I remember always struggling to find a single option that I felt fully represented me. I identify as both Black and Mixed—two of my grandparents were from Ghana, one was from Dominica, and one was from Lebanon. Was I "Black African"? "Black Caribbean"? "Arab"? I would usually end up ticking "Mixed"' and specifically "Any other Mixed or multiple ethnic background," but it would always feel unsatisfactory, as it didn't capture the fact that I am racialized as Black, which I think has a significant impact on the way I experience the world.

Lewis Recio, the legal and policy officer in Brussels, told me he also lived in the UK in the past, where he said he remembered experiencing a similar predicament whenever he had to complete forms asking about ethnicity. He thinks that the collection of race and ethnicity data in the UK and elsewhere could be improved by allowing people to tick all the options that apply to them, or even to describe how they identify in their own words. Both of these options would probably make analysis of the data more onerous, but gathering more granular data on how people identify could help to provide a clearer picture of how people's unique identities shape their experiences—including their health. Because they are socially constructed, racial and ethnic categories also tend to vary with time and place, with people shifting how they identify or the terms they use to describe themselves. Demographics also shift over time. In both the UK and US, Mixed or Multiracial people are among the fastest-growing racial/ethnic demographics. "I think the best is to provide a really flexible way and to regularly renew or adjust these categorizations," said Lewis Recio.[220]

Another limitation of race- and ethnicity-disaggregated data is that they often aren't disaggregated—that is to say, broken down—*enough*. When this is the case, the data can conceal inequalities that exist between smaller subgroups within populations. Tina J. Kauh, a senior program officer at the Robert Wood Johnson Foundation, a nonprofit organization in New Jersey, highlighted this issue in a 2021 paper, in which she and her colleagues called for further disaggregation of health data in the US beyond the commonly used broad racial and ethnic categorizations of "White or Caucasian; Black or African American; Latino or Hispanic; Asian American; Native Hawaiian and Pacific Islander; or American Indian and Alaska Native." Each of these categories encompasses a huge amount of diversity in terms of people's experiences—something Kauh became aware of growing up in the US as a child of Korean immigrants. "There's a model minority stereotype for Asian Americans, that they're thriving, they're doing really well," Kauh told me. "But Asian Americans are so diverse," she pointed out. "I personally saw firsthand the hardships that my parents experienced, that their social networks experienced," she said.[221]

When it comes to health, data disaggregation has helped researchers to see beyond racist stereotypes and identify inequalities between subgroups of Asian Americans, Kauh explained, such as disparities in health insurance coverage. According to an analysis of 2015 American Community Survey microdata by AAPI Data, an organization that publishes demographic data and policy research on Asian Americans, Native Hawaiians, and Pacific Islanders, 22 percent of Nepalese Americans and 20 percent of Korean Americans don't have health insurance compared to 6 percent of Japanese Americans. These and other disparities are hidden when data aren't disaggregated beyond broad classifications such as "Asian American"; analysis of the same data showed that the uninsured percentage among Asian Americans as a whole is 13 percent, compared to 9 percent among White, 15 percent among Native Hawaiian and Pacific Islander, 16 percent among Black, and 26 percent among Latino people. [222]

Lack of finer data disaggregation also disproportionately masks the experiences of smaller populations, such as Indigenous peoples. In cases where populations are particularly small, disaggregated data are often withheld due to privacy concerns or not collected in the first place, Kauh told me, which can present a barrier to addressing health inequities. "It's essentially wiping out entire tribal nations [from the data]," she said. During the first wave of the US's Covid-19 epidemic, for instance, an analysis by *The Guardian* found that almost half of state health departments that released demographic data on the toll of the virus failed to explicitly include Native Americans in their breakdowns, let alone tribal affiliations. "With what we're seeing right now, there will absolutely be a gross undercount of the effects of Covid on American Indian and Alaska Native peoples," Abigail Echo-Hawk, director at the Urban Indian Health Institute and chief research officer at the Seattle Indian Health Board, told the newspaper at the time.[223]

"The decisions around who we collect data on are often rooted in a system that is racist already," said Kauh. "What's really important is that we identify that balance of wanting to ensure data privacy and confidentiality without further making certain populations invisible," she told me.

Collecting disaggregated data on the health of populations in an ethical and equitable way is challenging, but not impossible. It requires investment in data collection and analysis, as well as engagement with communities to identify the right balance between representation of diverse populations, protection of people's privacy, and feasibility of data interpretation. That balance will shift depending on the specific needs and circumstances of populations. The goals of such data gathering should always be made transparent to ensure informed consent, and to make clear the important distinction between collecting information about race or ethnicity in order to measure the health impacts of racism, versus incorrectly using "race" as a biological category upon which to base medical testing or treatment decisions. Critically, data on health disparities must then be used to inform policies that address these inequalities.

It is also important to acknowledge that while efforts to improve the disaggregation of health data are certainly worthwhile, this alone won't be enough to close the racial and ethnic health gaps that we have been examining. That is because in addition to vast racial, ethnic, and caste health gaps within societies, there are also glaring data gaps in medical research. These gaps extend beyond health data, which we have just covered, to medical studies—from basic scientific studies of the mechanisms underpinning diseases, to large trials that test the drugs or other interventions that may one day treat or prevent them. Indeed, even though people of color often experience poorer health outcomes on average compared with White people, data points from non-White people are largely missing from medical studies. As of 2018, 78 percent of participants in genome-wide association studies—which aim to identify crucial connections between genes and disease—were of recent European descent. Clinical trial participants are also overwhelmingly White. This lack of diversity in medical studies means that insights derived from them may disproportionately leave people of color behind, potentially exacerbating existing health inequities. The failure to represent the breadth of humanity in medical studies is also stifling scientific discovery—and that is bad news for everyone, regardless of race or ethnicity.

On February 15, 2021, a judge in Hawaii ordered the pharmaceutical companies Bristol Myers Squibb and Sanofi to pay more than $834 million to the state for failing to properly warn doctors and patients that their blood-thinner drug clopidogrel, sold under the trade name Plavix, could be less effective among people of East Asian and Pacific Island ancestry. Judge Dean Ochiai in Honolulu ruled that the companies engaged in unfair and deceptive business practices between 1998 and 2010, by failing to add an adequate warning to the drug's label despite being aware of some of the risks. A number of studies have suggested that a genetic predisposition common among people of East Asian or Pacific Island descent results in poor metabolism of the drug, potentially leading to negative effects.[224]

But wait a minute. If race and ethnicity are social constructions, then how can a drug be racist? The short answer is that drugs can't be racist, but people and societies *can*. For the longer answer, we need to get into some genetics.

The issue with clopidogrel is largely centered around a gene called *CYP2C19*, which encodes an enzyme that is involved in metabolism of the drug. CYP is short for cytochrome P-450, and CYP enzymes are involved in the metabolism of many different drugs in the body. Genes, of course, vary between people—and that includes the genes that encode CYP enzymes. Natural variation in genes between people means that some people will have a version of gene X that metabolizes drug Y extremely well. On the other hand, some people will probably have a version that metabolizes drug Y a lot less well, and still others will likely fall somewhere in the middle. If we want to ensure that this variation isn't so significant that our imaginary drug would be ineffective for a significant proportion of people, then we would be best off testing our drug in a genetically diverse population. If instead we only tested our drug in a group of people from an arbitrarily selected subset of the population—people with blue eyes, for instance, or people who live in a specific village—we might inadvertently introduce a bias that means we fail to fully capture the diversity of responses to the drug that are present in the wider population. And that's *exactly* what we are doing time and time again in clinical trials. One study of trials conducted between 2001 and 2010 in the US found that about 83 percent of participants were White. The same study reported that between 1990 and 2000, that figure was 89 percent.[225]

While race and ethnicity have no biological or genetic basis, focusing disproportionately on an arbitrary subset of the population (such as White people) reduces the overall diversity among people included in medical studies and, critically, decreases the likelihood that insights derived from them will be relevant to the entire population.

It turns out that some people have a version of *CYP2C19* that leaves them unable to metabolize clopidogrel properly. This lowers the effectiveness of the drug, which is prescribed to prevent heart attacks and strokes in people at high risk of these conditions. If we again group people arbitrarily, for instance by racial or ethnic background, it is likely that we will see variation in metabolism of the drug between those groups. This is, indeed, exactly what we see in practice. For instance, one study reported that common variations in the *CYP2C19* gene seen in approximately 30 percent of people identified as White, 40 percent of people identified as Black, and more than 55 percent of people identified as East Asian significantly diminish their ability to metabolize clopidogrel. Another study found that 57 percent of Māori and Pacific Islander patients on the drug showed signs of low effectiveness of the treatment.[226]

Increasing diversity in drug trials will increase the probability that we pick up effects that might by chance occur at different levels within different subgroups of the population. It could also help us pick up effects that occur not by chance, but due to racism in societies. That fact isn't lost on Warren Whyte, who works in the biomedical industry in the US. Whyte is vice president of scientific partnerships at ConcertAI, a biotechnology company headquartered in Cambridge, Massachusetts. He is passionate about tackling racial and ethnic health inequities, particularly in cancer, and believes that improving representation of minoritized racial and ethnic groups in clinical trials will be an important part of that. We spoke in early December 2021, and he explained why.

"Our population is very diverse," Whyte began. "An African American does not experience the same quality of life and the same exposure to certain environmental factors as their White counterparts," he told me, adding that conscious and unconscious biases in medicine also contribute to racial and ethnic disparities in health. "It's not fair to assume that if you have a clinical trial that is just a White population or a single racial population, that that's going to somehow be reflective of the entire population," Whyte argued. "Because, while our DNA may be 99.9 percent the same across groups, unfortunately our lifestyles are not," he pointed out. "That definitely needs to be taken into account." If drug trials aren't representative of the entire population, they run the risk of producing therapies that don't benefit everyone equally, warned Whyte. After our conversation, my mind drifted back to clopidogrel. My hunch is that Bristol Myers Squibb and Sanofi might have been better off had they tested their drug among a more diverse group of people. After the Hawaii court ruling, both companies said that they would appeal the decision, which they claimed was "unsupported by the law and at odds with the evidence at trial."

The clopidogrel debacle is just one example of the way in which the lack of diversity in medical studies may result in therapies, diagnostic tests, and medical devices that work less well for significant proportions of the population—often people of color. For instance, if your doctor thinks you might have type 2 diabetes, they will want to check your average blood sugar level—something that typically means taking a glycated hemoglobin test. This diagnostic method is recommended by the WHO and is widely used around the world. But there's evidence that a gene variant present in almost a quarter of people with recent African ancestry alters the levels of glycated hemoglobin in the blood independent of blood sugar—meaning they may be more likely to be falsely diagnosed with diabetes. There is a similar issue in the case of cystic fibrosis—a condition that is underdiagnosed among people of African descent. Tests often look for known mutations within the *CFTR* gene, such as a mutation called deltaF508 that is present in 70 percent of people of European ancestry who have the condition. But among African descent populations, that specific mutation accounts for just 29 percent of cystic fibrosis cases. Instead, the cause is often one of a number of other possible mutations in the same gene—markers that may be less likely to be spotted since most investigations undertaken to identify *CFTR* gene mutations have been conducted in European ancestry populations.[227]

The problem extends to medical devices too. There is evidence that pulse oximeters—devices routinely used to test blood oxygen levels—work less well in people with darker skin. Pulse oximeters are usually attached to a person's finger, toe, or earlobe. They work by sending a beam of light through that part of the body and taking advantage of differences in light absorption between oxygenated and deoxygenated blood. They are considered so easy to use that patients have increasingly been self-administering these blood oxygen tests at home, particularly during the Covid-19 pandemic. But a study as early as 1990 indicated that inaccurate pulse oximetry readings were at least twice as common among Black patients compared to White patients. Subsequent trials have revealed significant skin pigment–related differences in the effectiveness of these devices, which tend therefore to overestimate the percentage of oxygen in the blood among darker-skinned people. This could have serious implications for patients, according to a 2021 report by the NHS Race and Health Observatory, an independent expert body funded by NHS England. That report was led by Olamide Dada, the UK-based doctor and founder of Melanin Medics. It called for an urgent review of pulse oximetry medical products used across the UK.[228]

"The research that we identified within that report was nothing new," Dada told me. "We went as far back as the 1990s stating what some of the research had suggested then," she pointed out. "But the relevance now is the [Covid-19] pandemic," she said. Since pulse oximetry is frequently used to guide clinical management of Covid-19, among other health conditions, Dada suspects that the lower accuracy of the devices among people with darker skin may be partly responsible for observed racial and ethnic disparities in Covid-19 outcomes observed in the UK and beyond. Following the publication of her report in March 2021, the UK's medicines regulator (MHRA) updated its guidance on the use of pulse oximetry, adding mention of the fact that darker skin pigmentation "may cause an overestimate" of blood oxygen readings. More recently, in June 2022, the UK government published plans to strengthen the regulation of medical devices, with a commitment that the MHRA would provide extended guidance on how manufacturers "can demonstrate and ensure the safety and efficacy of their products across diverse populations." To Dada, this represents a small, long-overdue step in the right direction. "This is so, so important, not just for the pandemic, but moving forward, with any medical device—why are we approving devices that aren't suitable to all groups?" she said.[229]

In genetic studies in particular, bias toward people of European descent additionally threatens to exclude underrepresented populations from the much-vaunted promise of personalized medicine. Predicting a person's risk of developing a particular condition based on their genome sequence is by no means a perfect science, but there is already evidence that it works far less well in populations traditionally underrepresented in genome sequencing studies. For instance, a 2019 study found that the accuracy of such disease predictions is approximately two times lower in Asian-descent populations compared to those with European ancestry, and about five times lower in populations of African descent. The authors of that study warned in their paper that clinical use of these polygenic risk scores, as they are called, could exacerbate existing health disparities.[230]

Even the way in which geneticists define the human populations they study is skewed. I am guilty too. In the last few paragraphs, I have been referring rather vaguely to "European descent," "Asian descent," and "African descent"' populations—language frequently used by geneticists. But these designations are ill-defined, overlapping, and imbalanced in terms of describing actual genetic variation between people. Referring to European and African-descent populations in the same breath, for example, might give the impression that these populations are somehow equivalent in terms of genetic diversity, when this isn't the case at all.

Because of humanity's common origin in Africa, where most human genetic development took place, populations with recent African ancestry harbor far more genetic variation compared with European descent populations. That doesn't mean that every single African person is wildly biologically different from every single European person. Rather, it means that within an African population, statistically you would expect to find a larger variety of genetic variants—versions of genes—compared to within a European population. In other words, it isn't the case that every single European person has example gene A, and every single African person has example gene B.

A better (although still imperfect) analogy would be something like this: if within a European population you have two commonly observed versions of a gene, A and B, then in an equivalent African population you might expect to have more versions of that same gene, including versions A and B, but also including versions C, D, E, F, and G. If we randomly pick two people from the European population then, there is a higher chance that they will both have the same version of the gene, because there are just two possibilities. If we randomly choose two people from the African population, on the other hand, we are more likely to find two people with different versions of the gene, since there are several possibilities. But it isn't as if the African population and the European population are two completely unrelated or isolated groups.

Humans have historically moved around and had sex a lot. This is reflected in human population genetics, which is much more of a continuum of genetic variation rather than something that can be carved into distinct geographic groups. That fact becomes obvious when we look at regions of the world where there has historically been a lot of intermixing between people from different continents, such as the Middle East. As Dorothy Roberts puts it in her book *Fatal Invention*, "If researchers collected DNA samples continuously from region to region throughout the world, they would find it impossible to infer neat boundaries between large geographical groups." Consequently, the groupings I have been using—and which population geneticists often use—are arbitrary in a biological sense; we could draw the boundaries between groups anywhere. However, they certainly aren't arbitrary in a social sense, given the extent to which they are racialized both in society and within medical research. That means we can't ignore them. And regardless of whether or where we draw boundaries between people, the fact will remain that including more people from different parts of the world in medical research will increase our understanding of the continuous spectrum of variation that is humanity—something from which we all stand to benefit.

Over the last few years, I have gotten to know several of a growing number of researchers and entrepreneurs around the world who have dedicated their careers to doing exactly that. Before we dive into the vast scientific benefits that increased diversity in medical research could offer, though, I want to highlight another, arguably more important reason why I think diversity among participants in medical research matters.

The underrepresentation of large swaths of humanity in medical studies doesn't only introduce selection biases and hold back potential scientific discovery: it also excludes many individual people from the benefits that can come from being a research participant. For patients who are terminally ill, access to new and emerging therapies through participation in a clinical trial might be their only hope to live; those belonging to groups that are frequently underrepresented in clinical studies may be more likely to miss out on these sorts of opportunities. Even just proximity to clinical research appears to carry benefits, without the need for direct participation. There is evidence from the UK, for instance, that people who receive health care in research-active institutions tend to have better outcomes and experiences of care than those treated in non-research settings. Similarly, research from the US suggests that hospitals that participate in clinical trials have lower mortality rates compared to hospitals that don't. Yet, as Warren Whyte explained to me when we spoke, these hospitals tend to be located in Whiter, more affluent neighborhoods, restricting access to the potential benefits of clinical research for communities of color situated farther away. "The distance that those people will have to travel, it may be too much for them, or they're going to go to a place that they've never been before to receive their care. So, that could be a challenge for those individuals," Whyte noted.[231]

Another barrier to research participation is preexisting health: people of color and those belonging to marginalized groups are more likely to be in poorer health, for all of the reasons we have been exploring in this book so far, which may disqualify them from participating in trials. Mistrust of the medical establishment among marginalized groups, due to personal experiences of discrimination in health care or as a result of the long history of racism and exploitation within medical research (more on this later), also presents a barrier to recruiting clinical trial participants from diverse backgrounds. None of this is helped by the lack of diversity among those leading the research. In the US, as well as in the UK, where I worked in academic medical research for several years, Black scientists like me were and still are woefully underrepresented in the scientific workforce and particularly in senior research positions. To address biases in medical research we must also address biases in *who* is conducting said research.[232]

Let's fast-forward to the future of health care. Artificial intelligence is guiding health care providers as they navigate complex decisions about patient care and—in combination with other technologies, such as wearable devices—enabling people to monitor and improve their own health, reducing their chances of becoming ill in the first place. That future is in fact already here, but it isn't as rosy as it might seem. Far from being neutral tools that can objectively guide diagnoses and decision-making, algorithms that are designed by humans and artificial intelligence systems are often trained using data from the very messy, very human, very *racist* real world. Existing racial inequities and biases are already being automated by these technologies, locking them into current and future health care.

It doesn't have to be this way. Ziad Obermeyer at the University of California, Berkeley, is one of a large number of researchers working to expose and address biases in AI. Obermeyer, who is also a trained physician, focuses his research on the intersection of machine learning and health. In 2019, a study he led revealed that millions of Black people in the US were missing out on health care because of racial bias in a widely used algorithm. The algorithm in question used data from people's past health records to predict the cost of their future health care. Based on the estimated cost, each person was assigned a risk score, which hospitals and some health insurance providers could then use to identify people who would be more likely to need more care in the future. The idea was that those with higher risk scores could then be prioritized for extra care to prevent them from becoming more unwell. But the study by Obermeyer and his colleagues found that Black people assigned the same risk scores as White people went on to have poorer health outcomes. That is because Black patients tend to generate lower costs than White patients for a variety of reasons (namely, racism and its negative effects on health—many of which we have already covered). But importantly, this doesn't mean they are actually less *sick* than White patients assigned the same score. Unfortunately, the algorithm wasn't developed in a way that accounted for this; as a result, the risk scores it generated resulted in White patients being prioritized ahead of sicker Black patients.[233]

Since they uncovered this inequity, Obermeyer and his team have been working with the company who developed the algorithm—as well as other companies and institutions who make and use similar ones—to eliminate the observed racial bias. "Before we published our paper, we reached out to the particular company that makes the algorithm that we studied and we alerted them to these findings," Obermeyer told me. "We actually worked with their technical teams and basically came up with a solution and retrained the algorithm with their own data and made it far less biased," he said, explaining that they achieved this by shifting the algorithm's focus away from cost and toward actual health outcomes.

Obermeyer and I spoke more than three years after he and his team's study was published, and it was clear that their work hadn't ended there. "We got a lot of publicity for that original study and so we got a lot of requests from insurance companies and hospitals that were using these algorithms and a bunch of other algorithms, that wanted to basically audit their own algorithms and find problems with them and fix them," said Obermeyer. As he explained to me, the algorithm that he and his colleagues analyzed in their study is just one among a whole family of similar health prioritization algorithms developed and used by lots of different companies and institutions, including the US government. "They're actually trying to do something that algorithms are, I think, very promising for, which is to—in a population of patients—figure out who's going to get sick later, so that we can target help to them today," he said. "But we have to be so careful when we're building those algorithms," Obermeyer cautioned. He and his colleagues have been working with some of the developers of these sorts of algorithms to root out biases—and, while he acknowledges the fact that many algorithm developers have taken the initiative to tackle these problems, he thinks there should be more formal regulation on the use of this technology. "I've been working in the US with a few different regulators at the national level, but also law enforcement officials at the state level, on how to do investigations, how to audit algorithms, and how to ensure accountability," he told me.[234]

Any future auditors will have their work cut out for them, because AI is already widely prevalent in health care technologies and its use is on the rise. The health care AI market is expected to be worth more than $34.5 billion by 2027 in the US alone. Large technology companies are keen to get a slice of the action. Google's parent company Alphabet is particularly ambitious; about halfway through 2022 it had already invested $1.7 billion into futuristic health ventures. But some of its health technologies have garnered attention for the wrong reasons.[235]

In 2019, the US health news website STAT reported that the technology used in wearable heart rate trackers made by Fitbit, since acquired by Alphabet, could be less reliable for people with darker skin. Two years later, a dermatology app developed by Google researchers to help identify skin conditions from photographs came under scrutiny for a similar reason. In an online statement announcing the app in May 2021, the company boasted that it was able to recognize 288 different skin conditions from photographs. "To make sure we're building for everyone, our model accounts for factors like age, sex, race and skin types—from pale skin that does not tan to brown skin that rarely burns," the statement also noted. But a scientific paper published by the research team a year earlier, demonstrating the efficacy of the approach used by the app, revealed that the data used to train the algorithm were skewed toward lighter skin tones.[236]

During my conversation with Carlos Charles, the New York–based dermatologist we met earlier, he told me that he worries about skin color biases being transferred from medical education images straight into algorithms without a second thought. "It makes me nervous," said

Charles. "Textbooks are biased," he noted. "And then if we make technology biased too, it's going to be very problematic," he pointed out. When we spoke in January 2022, Charles told me he was aware from communication with Google that the company was trying to better incorporate darker skin tones into its algorithm. "But I don't know to what extent," he said. In May 2021, after Google announced its app, a spokesperson from the company told VICE Media's technology magazine *Motherboard* that the entire field of dermatology suffers from a lack of data and images of non-White patients and that accounting for that problem was a central concern for its researchers as they designed the dermatology app, which would be further developed prior to public release.[237]

Having spent years learning about the multiple ways racism harms people's health, including countless examples of racial biases within health care itself, the idea of all those prejudices being automated through AI was sending chills down my spine. Obermeyer's optimism on the subject, though, surprised me—and gave me hope. He believes passionately that AI can be a force for good. In fact, he thinks that AI may hold the key to identifying and tackling human biases in health care. In a sense, this is precisely what the health care prioritization algorithm that his team analyzed in their study did. It exposed the fact that Black people in the US generate lower health care costs, which is a product of racism. This is a common theme. Another study by Obermeyer and his team revealed racial bias in an algorithm used by the US government to allocate Covid-19 relief funding to hospitals. The government's formula distributed funding from a $175 billion Covid-19 relief package to hospitals based on their revenue, leaving already underfunded hospitals that disproportionately serve Black patients with less money to manage the larger volume of Covid-19 cases that they were seeing. Obermeyer and his colleagues want to harness the power of AI to expose these sorts of biases *before* they get incorporated into widely used algorithms. In a 2021 study, they put this idea to the test.[238]

Obermeyer and his team used machine learning to train an algorithm to identify knee pain, using a collection of data from more than four thousand osteoarthritis patients in the US. But their approach was slightly unusual. "The way most machine learning papers work is they put in a bunch of X-rays and then they train an algorithm to learn from what a doctor would say about the X-ray," Obermeyer told me. "So, is there a lot of arthritis in this knee or not?" But in addition to feeding the algorithm with information from knee X-rays and doctors' interpretations of them, Obermeyer and his colleagues also provided it with information that came directly from the patient, regarding their perceptions of pain. "We trained the algorithm not just to learn from the doctor, but also to listen to the patient and identify which knees are going to be painful," he explained.

The results were striking. "The algorithm did a much better job of identifying which knees were going to be painful than the radiologist," said Obermeyer. "So, it was identifying a lot of signal that the radiologist was missing for explaining the pain that a patient was reporting." And this was particularly beneficial for Black patients, who tend to score significantly higher on knee pain scales compared to White patients. "There's this huge difference in pain between Black and White patients whose knees look the same to the radiologists, where Black patients report a lot more pain," said Obermeyer. He thinks this reflects biases in the way health care providers usually assess knee pain. "The grading system that we use today to judge the severity of arthritis in someone's knee—that was built up from studies of coal miners in Lancashire [UK] in the 1940s and '50s," Obermeyer explained to me. "Of course, populations today are very different from that original population. But the knowledge hasn't been refreshed," he pointed out. "I think there's this very appealing way in which algorithms can provide this objective refresh of medical knowledge and can actually discover new things that we don't know about yet," he added. In their study, Obermeyer and his colleagues found that, compared to the status quo, use of their algorithm would have doubled the proportion of Black patients eligible for knee replacement surgery from 11 to 22 percent.

"Algorithms and data in general reflect the world as it is," Obermeyer told me. "Not the world as it should be." It is up to all of us to make our world more equitable, including in the context of health. Data and algorithms can be powerful tools—something the medical research world is increasingly appreciating, as we will discover in the next chapter. But we should wield these tools carefully, with oversight and accountability, to avoid perpetuating existing harms.

CHAPTER 9

Closing the Data Gaps

In 2016, Frank Brown became part of a medical success story. A business owner and grandfather of six based in Dallas, Texas, he had been experiencing dangerously high cholesterol levels that weren't responding to conventional treatments. It was the result of familial hypercholesterolemia, an inherited condition that causes high levels of cholesterol—particularly of a form called LDL cholesterol, which is sometimes referred to as "bad cholesterol" because of its association with cardiovascular disease. Frank had already had two heart attacks, and despite trying various cholesterol-lowering medications, his cholesterol levels remained persistently high. Fortunately for him, a study led by geneticists Jonathan Cohen and Helen Hobbs at the University of Texas Southwestern Medical Center, also based in Dallas, had yielded an intriguing finding years earlier that would eventually form the basis for the development of a new kind of treatment to help him and patients like him.[239]

In their study, Cohen and Hobbs analyzed DNA sequences from 128 people with naturally low levels of LDL cholesterol—essentially the opposite of Frank's situation. They discovered two previously unknown variants of a gene called *PCSK9*, which were linked with particularly low LDL cholesterol levels. The identification of these gene variants paved the way for the development of two drugs that block PCSK9 (the protein encoded by the *PCSK9* gene), mimicking the biology of those people possessing variants of the gene that naturally reduce LDL cholesterol. These PCSK9 inhibitors, as the drugs are collectively known, have been able to help patients like Frank to better control their cholesterol. Frank first received one of the two drugs as part of a clinical trial, and they have since been approved by the FDA.

"There are many more Frank Browns out there—patients who can't control their cholesterol with the standard drugs," said Frank's cardiologist, Amit Khera, in a statement published by the University of Texas Southwestern Medical Center in 2016. "It's wonderful to have this option to offer this special set of patients."[240]

The study by Cohen and Hobbs also demonstrated something else: the power of diversity for genetic discovery. Their study population was derived from the multi-ethnic Dallas Heart Study, and about half of the study participants were people with recent African ancestry. It was thanks to this diverse study population, including a significant proportion of people of African descent, that the researchers were able to identify those previously undiscovered variants of the *PCSK9* gene. According to their analysis, those variants happened to be rare among the study participants of European descent, occurring at a combined frequency of less than 0.1 percent. However, they were more common among those with recent African ancestry, occurring at around 2 percent and providing enough data points for the researchers to confirm the association with low LDL cholesterol. If, like most human genetic analyses, Cohen's and Hobbs's study had focused predominantly on people with mainly European ancestry, it is unlikely they would have been able to make the discovery that they did.

The PCKS9 discovery is just one example of the benefits that increasing diversity among medical study participants can bring. We know that data gaps may be contributing to health inequities. The good news is that in recent years there have been increasing efforts by scientists to change that. Among these scientists is Deepti Gurdasani, a clinical epidemiologist and statistical geneticist at Queen Mary University of London.

I first interviewed Gurdasani in October 2019, a few days before a study she had co-led in Uganda was going to be published in the scientific journal *Cell.* I had just started an internship at *New Scientist* in London, having recently completed my PhD, and I had only authored a handful of articles in the magazine so far, so I was delighted when I got the opportunity to speak to Gurdasani and report on the research she had been involved in. She, along with a huge international team of scientists spanning Africa, Europe, and North America, had collected and analyzed DNA sequences from more than six thousand people across twenty-five villages in south-west Uganda—including almost two thousand whole genomes—producing one of the largest DNA datasets from an African population. Gurdasani and her colleagues also gathered information about the health of the participants in their study and incorporated additional data from fourteen thousand people from other countries, including Ghana, Kenya, Nigeria, and South Africa, as well as Uganda, into their analysis. They uncovered a huge amount of genetic diversity, about a quarter of which hadn't been recorded before. Of particular interest for medical research, they identified 41 gene variants associated with cardiovascular and metabolic health, 23 of which hadn't previously been discovered.[241]

The reason that so much of this genetic variation had remained hidden to science until now, Gurdasani explained to me, was because most human genomics studies focus overwhelmingly on populations of European descent at the expense of other populations. In fact, genomics study participants are disproportionately recruited from a tiny subset of countries globally. One analysis reported that 72 percent of genetic discoveries made between 2005 and 2018 were derived from study participants recruited from just three countries: the US, the UK, and Iceland.* Another found that 78 percent of participants included in genome-wide association studies, which search for links between genes

* Iceland has been a popular destination for human genetics research, due to the perceived relative genetic homogeneity of its population (although some geneticists have questioned this notion of homogeneity).

and disease, were of European descent. "If we continue to sample Europeans and extend our findings to other populations, then that certainly is not going to work for everyone," Gurdasani commented, when we spoke again a few months later. Indeed, among the findings in her and her colleagues' paper was the discovery that a gene variant found among 22 percent of people with recent African ancestry alters levels of glycated hemoglobin independent of blood sugar, potentially reducing the effectiveness of a widely used diabetes test for these individuals.[242]

Including a wider range of people in genomics studies isn't only a matter of fairness—it makes scientific sense. Gurdasani believes that increasing diversity in genomics research holds the key to unlocking new therapies for all manner of health conditions—therapies from which we could all benefit.

The ability to sequence entire human genomes has transformed medicine. That transformation began with the Human Genome Project, completed in 2003, which provided us with the (almost complete) genetic map of a human for the first time.* It led former US president Bill Clinton to famously—and quite rightly—quip that "all human beings, regardless of race, are more than 99.9 percent the same." Genetics, however, is the study of *difference* and when it comes to human genetics there has been significant interest from researchers in harnessing the tiny proportion of DNA that *differs* or varies between people to try to answer medically useful questions such as *Why do some people get a particular disease while others don't?*[243]

Since the first human genome was sequenced, whole-genome sequencing has become much quicker and less costly, paving the way for so-called genome-wide association studies or GWAS. These studies were borne out of an increasing recognition among geneticists that most human diseases are polygenic, meaning they are influenced by multiple different genes, and often also by environmental factors. Genome-wide

---

* The Human Genome Project was considered complete in April 2003, but only covered about 92 percent of the total human genome sequence. The first complete sequence of a human genome was published in 2022.

association studies aim to identify these various genes by comparing whole genome sequences from a group of people who have a particular disease, with those from another group of people who don't have that disease. The logic is that gene variants (sometimes referred to by geneticists as single nucleotide polymorphisms or SNPs) that are more common in the disease group than in the non-disease group, are more likely to be associated with the disease. This approach has improved our understanding of countless health conditions and has contributed to the identification of new therapeutic strategies for several, such as arthritis and inflammatory bowel disease. And it benefits from diversity: the more genetic variation there is within the population of people whose DNA is being analyzed, the greater the probability of identifying those key genetic differences between the "disease group" and the "non-disease group." As a result, neglecting to comprehensively capture the diversity of humanity in genomics studies is constraining precisely this kind of scientific discovery. From the perspective of geneticists, such as Gurdasani, this means that we are collectively missing out on the huge untapped potential that greater inclusion has to offer for the possibility of new genetic discoveries.[244]

The vast underrepresentation of African-ancestry populations in particular has been especially limiting for genomic discovery. A 2018 analysis of genome-wide association studies found that although only around 2 percent of the subjects included were of recent African descent, their genetic information contributed to 7 percent of associations—those vital links between genes and disease that can help inform treatment development. The same analysis found that people identified as Latin American—a broad designation that encompasses many people with various ancestries, often including African, Indigenous American, European, or some combination of these—accounted for only about 1 percent of studied samples but contributed to approximately 4 percent of associations. The opposite trend was seen in populations of predominantly European descent, which accounted for 78 percent of samples studied yet only 54 percent of associations.[245]

These findings aren't surprising. As Gurdasani put it: "European populations are certainly not the most diverse populations in the world. Humans originated in Africa, and then a subset of individuals migrated out of Africa and populated the rest of the world." As a result—and as I have mentioned—genetic diversity within Africa far exceeds that in the rest of the world. "And the more diverse populations you study, the more opportunities you have to identify associations with disease, which is what leads us to targets for drugs and new therapies," Gurdasani explained. What's more, those therapies will be more likely to work for everyone.

Fortunately, the world is waking up to the glaring data gaps that exist in medical research, as well as to the value that increased inclusion has to offer. Researchers are actively endeavoring to include more participants from traditionally underrepresented groups in genomics studies—a goal supported by initiatives such as Human Heredity and Health in Africa (H3Africa), GenomeAsia 100K, and the US National Institutes of Health's All of Us Research Program—and many are already reaping the benefits in terms of new genetic discoveries. The findings made by Gurdasani and her colleagues through their Uganda-based study, supported in part by the H3Africa initiative that aims to boost genomics research on the African continent, represent just one recent example.

The H3Africa consortium—funded by the US National Institutes of Health, the Wellcome Trust, and the African Academy of Sciences—includes forty-eight research projects based in thirty-four countries across Africa. In October 2020, a genomics study led by researchers in the consortium, including Zané Lombard at the University of Witwatersrand in South Africa, reported the identification of millions of never before discovered gene variants. Lombard and her colleagues sequenced the whole genomes of 426 people from thirteen countries in Africa: Benin, Botswana, Burkina Faso, Cameroon, Democratic Republic of the Congo, Ivory Coast, Ghana, Guinea, Mali, Nigeria, South Africa, Uganda, and Zambia. Collectively the study participants represented fifty different ethnolinguistic groups, making it one of the most diverse such investigations ever. And the results didn't disappoint. "We discovered 3 million new variants that nobody else previously had described before," Lombard told me. I caught up with her just over a year after she and her colleagues published their findings in the scientific journal *Nature*. She told me that the mountain of new genetic information they had uncovered had been a significant boon for her research, which focuses on rare diseases.[246]

When a child is born with a rare disease for which an underlying genetic cause isn't known, Lombard explained to me, geneticists like herself can use genome sequencing to try to identify the specific gene variant that's responsible for the condition. For instance, by comparing DNA from the child, who has the condition, with DNA from their parents, who don't have the condition, she could start to narrow down the list of possible gene variants that might be causing it. She would then typically compare this whittled-down list to a database of DNA sequences obtained from a larger population of healthy adults, in order to pinpoint any gene variants that are present in the child but rare in the wider population. The logic is that if a gene is common in the broader population, it is unlikely to be disease-causing. "So what we've seen is that some of those variants then would be classified as disease-causing, based on their frequency," Lombard told me.

But the lack of diversity in DNA databases poses a problem here too: the overrepresentation of samples from European-ancestry populations means that variants not seen often among Europeans are more likely to be classified as rare, and therefore as disease-causing. Lombard and her team's study revealed that these classifications aren't always correct. They found that several variants previously identified as rare and probably disease-causing were in fact common among the diverse participants included in their study. This information is extremely valuable for unraveling disease mechanisms, as well as for our understanding of the roles that various genes play in human biology.

"We can reclassify those variants and say, this is maybe very rare in European populations, but we see it at a higher frequency in this African population, and therefore you need to readjust your thinking about this variant—it's probably not disease-causing," said Lombard. In other words, being able to reclassify gene variants previously incorrectly identified as disease-causing helps geneticists to better distinguish the signal from the noise when it comes to homing in on the true causes of rare conditions. It enables them to cross some variants off their list, bringing them closer to discovering the actual gene or genes implicated in a given disease, Lombard explained.

The kind of personalized, genomics-based approach to medical diagnosis that Lombard described isn't yet widely available to patients around the world, but she and her colleagues are motivated to ensure that once it is, it will work equally well for everyone. "When we start thinking about how genomic data is going to be used in the future for things like precision medicine, it's really important that there is global representation," Lombard emphasized. "Already we're seeing this divide in applicability of the tools that are being developed from a genomics point of view, that are maybe not as useful or applicable to certain groups because the genetic diversity is not represented in the way that it was designed and informed," she pointed out. She and her colleagues are motivated to change that. "We do want to show that precision medicine can be applied to all global populations," she said.[247]

A few weeks after my conversation with Lombard, I got to know another geneticist who explained to me how increasing efforts by researchers, such as Lombard and Gurdasani, to diversify DNA databases were directly benefiting her research on the other side of the world. Maria Chahrour is based at the University of Texas Southwestern Medical Center (coincidentally the same research institution where Cohen and Hobbs conducted their study on genetics and cholesterol). Chahrour's research focuses on investigating the complex contribution of genetic factors to autism spectrum disorder, including through genome sequencing studies. In 2016, she had an ongoing study for which she was recruiting local families. "I really wanted to specifically focus on the diverse population in Dallas," she told me.

It was around this time that Chahrour met her soon-to-be collaborator Leah Seyoum-Tesfa, president of Reaching Families Advocacy and Support Group (REACH)—a nonprofit organization that supports East African immigrant families with disabled children. Seyoum-Tesfa is originally Ethiopian, and two of her four children are on the autism spectrum. Before she started her nonprofit in 2011, she worked as a nurse practitioner. Her master's thesis investigated the prevalence of autism among East African children in Texas, and when she heard about Chahrour's study, she was keen to get her community involved.

"When I met Leah, I was super excited," Chahrour told me. "The two of us clicked," she said. Chahrour visited the REACH community center several times and spoke with community members about her research on autism and genetics, before sharing more information about the study she was recruiting for at the time. There was a lot of interest in the study, but many people also had questions and concerns about participating, including about the use of their data and the stigma associated with autism. Chahrour and Seyoum-Tesfa worked together to incorporate changes to the study design, in collaboration with the community. Eventually, this evolved into a separate research project. "We basically started a separate study, specifically for the East African community," Chahrour explained. When I spoke with Chahrour, she and her colleagues were in the midst of analyzing the first results from that study.

At this point you probably won't be surprised to learn that research on the genetics of autism, much like genetics and genomics research more broadly, has predominantly focused on populations of European descent. Research conducted so far already indicates that there is a huge amount of diversity in terms of how autism manifests and the genes involved, Chahrour explained to me. Autism or autism spectrum disorder in fact refers to a wide range of conditions, influenced by a complex constellation of genetic and environmental factors. But being a geneticist, Chahrour is particularly interested in learning more about the genetic factors that contribute—something she hopes could help improve diagnosis. From a genetics standpoint, "you can think of autism as this big collection of individually rare diseases," she told me. "So, each of these genes that we find contributes to less than 1 percent of patients. And, in total, we've identified genetic causes for only about 35 percent of cases," she said. Chahrour believes that including more diverse populations in studies could help uncover as yet unknown gene variants that contribute to autism. Early findings from her team's study support this belief.

Chahrour and her team analyzed DNA sequences from 129 people, including 36 children with autism spectrum disorder and their families. They identified more than two million previously undiscovered gene variants, which they then began to pore over in search of variants that might be associated with the condition. To do this, they took a similar approach to the one Lombard described for identifying the genetic causes of rare conditions: they compared DNA sequences from the children with those from their parents, and then sought to compare them with those from a reference population of people without autism, all the while searching for rare variants that were unique to their study cohort of children with the condition. But the lack of diversity in most DNA databases posed a challenge in pinpointing the variants that were truly rare.

Fortunately, Chahrour was able to gain access to a more diverse database. "I was lucky enough to connect with someone from the H3Africa consortium, and I applied for data access," she told me. By comparing the DNA sequences from their study with about four hundred sequences from the diverse reference database, she and her colleagues were able to identify more than forty gene variants potentially associated with autism that haven't previously been linked to the condition. Their results were published in a scientific journal in 2023. Chahrour thinks her team's findings demonstrate the value of including diverse populations in research as well as boosting diversity in DNA databases.

"I think the genetics community as a whole recognizes this need for diversity and the need to do more sequencing on the African continent," she said. "And they are working to be more inclusive, I think, slowly." Chahrour and Seyoum-Tesfa have additionally started collaborating with researchers based in Ethiopia, where many of the participants in their study are originally from. "We're hopeful that as we continue to work together with the community and the families and other researchers and foundations that we will be able to advance the collective knowledge," said Chahrour.[248]

In addition to the increased focus on African populations, geneticists are also increasingly turning their attention to smaller, more isolated populations—including of Indigenous peoples both within and outside Africa—in the hope that analyzing their DNA might provide insights into unique environmental adaptations connected to health. Analysis of DNA sequences from Greenlandic Inuit populations, who have relatively low levels of heart disease despite eating fat-rich diets, has helped to yield new insights into the relationship between genes, diet, and heart health, for instance. And, more recently, an analysis of genomic data from Melanesian populations revealed new variation in several genes linked with metabolism.[249]

It isn't only academic scientists who are recognizing the power that diversity has to enhance medical research. Commercial genetic testing companies, such as 23andMe, are also seeking to diversify their DNA databases and big pharma is paying attention, too. For instance, in 2021, Novartis and GSK announced a collaboration to support research investigating links between genetic diversity across various regions in Africa and its potential impact on response to therapeutics. But it is a biotechnology start-up founded in Nigeria in 2019 that really caught my attention as I was researching this emerging industry. The stated mission of 54gene was, at the time, to equalize precision medicine by "pioneering the inclusion of the African genome in research" (it is named to reflect Africa's 54 countries). The company's cofounder and former

CEO Abasi Ene-Obong very much embodied this goal. The two of us spoke around the time of his company's three-year anniversary.[250]

Like Gurdasani, Lombard, Chahrour, and other geneticists I've spoken with about this topic over the last few years, Ene-Obong believes that diversifying the world's DNA databases holds the key to making medical research more equitable, and in particular to ensuring that the precision medicine of the future will work for all populations globally. Being Nigerian, he is especially passionate about ensuring that African people benefit from the rush by scientists to capitalize on their genomic data. "We don't believe in helicopter science," he told me when we spoke. "Where people come and swoop in, take the data, take the samples, and go do everything else outside of Africa," he said. "We have our values as a company and part of that is to promote the sciences in Africa. And so, we built our lab—or labs—within Africa," he explained.

Ene-Obong is fiercely proud not only of the company's African roots, but also of its early scientific achievements. "Prior to 54gene, no company was doing human whole genome sequencing in Africa," he told me. The company was also somewhat exceptional in its decision to pivot away from obtaining DNA samples from consumers by selling ancestry testing services, focusing instead on the medical aspect of its work. 54gene did initially start selling saliva sample kits, promising to provide people with insights about their ancestry and health, but it abandoned these quite quickly. A 2019 report about the company in *Wired* suggested this may have been partly influenced by distribution challenges in Africa, but Ene-Obong told the magazine it was also motivated by a desire to increase transparency. "We switched strategy," he said, when I asked him about the decision. "We decided that there was no need to use a Trojan horse business," he explained. 54gene instead obtained samples by partnering with hospitals and researchers, starting in Nigeria, enabling it to access patients' samples with their informed consent. "We want people who give us their data to know what they're giving us their data for," he said.[251]

When we spoke, however, Ene-Obong didn't rule out the possibility that the company could go back to selling consumer DNA tests again in future. "It wouldn't be because we're trying to use it to get data, it would be because we want to give back to our community," he assured me. "There's lots of people who are somewhat disenfranchised, because of slavery and colonialism," he said, adding that many people of African descent have a desire to learn more about their ancestry as a result of these histories. It is an urge I can relate to myself, although I am personally unconvinced that the answers to my identity crisis lie in DNA ancestry tests (those tests can't say anything about where in the world our ancestors lived, they can only tell us where in the world we are likely to have living relatives today). 54gene's main focus is health, though: when Abasi and I spoke it was in the midst of sequencing genomes and analyzing health data from one hundred thousand people in Nigeria, for starters. The company aimed to use these and other data it gathered from people across Africa to inform the development of new medicines.[252]

"I decided that rather than be a company that sold genetic data, that we should be a company that developed medicines," Ene-Obong told me. There is potential for a lot of money to be made by selling genetic data or selling drugs developed using such data. In 2019, for instance, UK Biobank—a charity supported by the UK government—sold exclusive access to genetic data from hundreds of thousands of participants in its UK-wide study to four of the world's biggest pharmaceutical companies in a deal worth approximately $123 million. Meanwhile, 23andMe is profiting similarly from its consumers' genetic information. GSK invested $300 million in the company in 2018 and paid separately for a four-year partnership granting it exclusive rights to use 23andMe's data to develop drugs (that partnership had been recently extended for another year at the time of writing). In 2020, 23andMe additionally sold the rights to a drug it developed based on its consumers' DNA to the Spanish pharmaceutical company Almirall.[253]

Ene-Obong believed that 54gene's diverse genomic database would give it an edge in informing the development of new medicines, and he would want any drugs developed this way to be accessible in Africa right away. Typically, he pointed out to me, "Drugs only come into the African continent when they are generic—when they're off-patent—so, that could be twenty years after these drugs were sold in the US." He said he hoped his company would be able to change that. "I want to make a difference in my homeland," he told me. "I want to make a difference in the world, but I want the difference to also be there in the continent [of Africa]. So that's why I'm doing this," he continued. "I remember when I was starting this and everybody was like, 'You know you could just turn it around and sell for half a billion, billion in the next one year.' But that's not why we're doing this. We're doing this because hopefully, if we are here, and we are the ones doing this, we can make a whole lot of difference for a whole lot of people," said Ene-Obong.

At the time I interviewed Ene-Obong, 54gene was just one among a number of companies that looked poised to profit from the growing appreciation for diversity in the health care and medical research industries. When we spoke, Warren Whyte, the vice president of scientific partnerships at US biotechnology firm ConcertAI, was leading an initiative launched in 2020 called Engaging Research to Achieve Cancer Care Equity or ERACE (pronounced like "erase," because the initiative aims to erase cancer disparities, Whyte explained to me). The initiative is focused on eliminating racial and ethnic inequities in cancer in the US by improving representation of racial and ethnic minorities in cancer-related clinical trials. A tool developed by ConcertAI, using anonymized data from nearly five million cancer patients across the country, can help clinical trial organizers to identify hospitals and health care providers who see a lot of patients belonging to traditionally underrepresented racial and ethnic groups, for instance, so that they can be better targeted for recruitment. The tool also recommends modifications to trial recruitment criteria, helping to flag ways in which organizers can enhance recruitment from underrepresented groups in their trials without substantially changing the designs of their studies. The ERACE initiative is currently focused in the US, Whyte told me, but ConcertAI aims to employ equivalent approaches globally, as similar racial and ethnic health inequities exist in other countries too. "This is not just a US problem, it is an international problem as well," said Whyte.

In July 2022, while attending an international conference about racism and health, I learned about another company focused on boosting inclusivity in clinical trials—this time in the UK. In a prerecorded video played to attendees of the Health, Race and Racism International Conference, organized by the NHS Race and Health Observatory, Ash Rishi, founder and director of COUCH Health, spoke about his personal motivation and vision for the company.

Rishi's parents immigrated to the UK from India in the 1960s and '70s, before they were wed in an arranged marriage. Rishi was their only child, and when he was a teenager, his father began experiencing a collection of unexplained symptoms that would only be recognized three years later as signs of prostate cancer. "We lived in Surrey in the nineties, which was a very White area, a very affluent area, so the health care teams would reflect the area," Rishi told the audience at the conference. "My father's diagnosis was missed because of biases," he added. In the three years leading up to the diagnosis, his father's health had deteriorated significantly. "The symptoms got worse—back trouble got worse, burning in the stomach, losing weight, losing hair," Rishi recalled. "And we tried all of our, you know, Asian remedies, homeopathic medicines, we tried changing his diet—it just got worse," he said. Eventually his father demanded that his family doctor refer him to a hospital, which is where he was finally diagnosed with prostate cancer—by which point the diagnosis was terminal. According to Rishi, the biases didn't end there. "One of the biases that stayed with me throughout my whole career was that my father's oncologist was actually an investigator for a global phase three clinical trial in prostate cancer. It transpires that my father wasn't actually eligible, because he was young and they wanted older individuals for their clinical trials, which always blows my mind, because prostate cancer is now—people are getting [prostate cancer] younger and younger and younger," he sighed.

Rishi couldn't help but wonder if he could have had more time with his father, had he been able to participate in that clinical trial. "He could have seen me pass my A-levels, he could have seen me go to university, he could have gone in a way that he would have been proud to see something of his only son's life," said Rishi. He and his mother, whose English wasn't fluent, struggled to support his father as he navigated the health care system during the final stages of his life. "We didn't have any family in the UK. We didn't have a support network that we could rely upon that could help us through that situation," Rishi explained. This was exacerbated by the fact that his father's end-of-life care wasn't culturally competent, he said. It was the combination of all of these experiences that drove Rishi to found COUCH Health, a patient engagement agency aiming to boost diversity in clinical trials and make health care more equitable. The company, based in Manchester, UK, conducts research with patient groups and caregivers regarding their unmet needs and supports clinical trial organizers to make their studies more accessible and inclusive. Rishi hopes this will help to increase opportunities for patients like his father to participate in clinical trials.

Aspirations to increase participation in genomics studies and clinical trials are laudable—and I have no doubt that the entrepreneurs and academic researchers who are driving efforts to achieve this are well-intentioned. But a growing number of researchers and activists are rightly concerned that the accelerating academic and commercial interest in diverse health data raises serious ethical issues. To tackle racism in medical research and in health more broadly, we must strive not just for inclusion—but for equity.

CHAPTER 10

The Illusion of Inclusion

As I learned more about growing efforts by researchers to increase inclusion in medical studies, and about the benefits of diversity for furthering scientific discovery, I thought back to the vast racial and ethnic health gaps that we have been examining throughout this book. While I believe that increasing diversity in medical research is a worthy goal that carries many potential benefits, I think it is important to put those benefits into perspective—and to ask whether everyone will benefit *equally*.

Let's take, for example, the goal of sequencing more DNA from traditionally underrepresented populations. We now know that increasing diversity in genomics studies enhances the potential for genetic discoveries, which can inform the development of new treatments for diseases. Similarly, increasing diversity in clinical trials could help ensure that any treatments that get licensed work well for diverse groups of people. At the same time, futuristic personalized or precision medicine approaches—which, as we saw, also benefit from more diverse DNA databases—could increasingly enable treatments tailored to people's individual needs. But these scientific and medical advances won't go very far in tackling racial and ethnic disparities in health if we don't deal with the major driver of these inequities.

A fancy new medical treatment won't be able to do much good for public health if it is inaccessible to many of the most disadvantaged people within societies and around the world, because of racism. This is a terrible irony, since racism—as we have discovered—is a huge contributor to poor health in the first place. And while harnessing the tiny amount of genetic variation that exists between people can be valuable for understanding disease processes and guiding drug discovery, it is worth remembering, firstly, that far more of that variation exists within geographic populations than between them and, secondly, that there are much more straightforward, nonmedical approaches that already exist to tackle racial and ethnic inequities in health. For example, we know perfectly well why Ella Kissi-Debrah in London spent much of her short life experiencing severe asthma—and why this eventually killed her at just nine years old. The reason—air pollution exposure—is quite literally mentioned on her death certificate (thanks in no small part to campaigning by her mother Rosamund), and as we learned earlier in this book, environmental racism means that exposure to air pollution and its well-established harmful health effects is unequal. We can conduct the most diverse medical studies in the world and develop all the possible drugs to cure every single kind of asthma, but if we don't tackle racism in all its forms, racial and ethnic health inequities will persist.

There's another problem: efforts to include more people from traditionally underrepresented groups in medical studies can risk exploiting the very same groups of people many of these studies are so desperately seeking to include. Keolu Fox, an anthropologist and genome scientist at the University of California, San Diego, is acutely aware of this potential pitfall. At the same time, the benefits of diversity for genomic discovery aren't lost on him. In fact, he has dedicated much of his career to advocating for increased diversity among genomics research participants, pointing to the potential scientific advantages this could bring. As he put it when I interviewed him in 2020: "There's this treasure trove of human genetic variation that could lead to a new understanding of human biology." But, when we spoke, Fox also warned of the danger that the growing rush to boost diversity in medical studies could result in further exploitation of already marginalized groups. "There's this new modality of treating Indigenous people's genomes like coal, cobalt, diamonds or oil," he told me. "It's the definition of colonial [. . .] It is exploitative and extractive," he said. And yet, Fox pointed out, "there's this illusion of inclusion."

Medical researchers at universities and companies should heed Fox's warning. Medicine has a dark history of exploiting marginalized groups, which has left many people understandably wary about participating in medical research. That there are benefits of increasing diversity in medical studies is clear—the challenge will be in ensuring that those benefits are spread equitably, and without causing further harm to historically (and currently) marginalized groups. Most of the geneticists I have spoken with about this agree—including Deepti Gurdasani, Zané Lombard, and Maria Chahrour, all of whom we met in the previous chapter.

During one of our discussions, Gurdasani pointed out to me that there is a long history of samples and data, taken from patients or research participants belonging to marginalized groups, being used without consent for other purposes than those agreed to or understood. Indeed, cells that I used for part of my PhD research—and which have been widely used in biomedical research laboratories around the world for decades—were originally derived in the 1950s from an African American woman named Henrietta Lacks, when scientists at Johns Hopkins Hospital took and propagated samples of her tumor without her or her family's consent. George Gey is the scientist credited with culturing her cells to produce the first ever immortalized human cell line—named HeLa, after Henrietta's name—which has since enabled numerous scientific medical breakthroughs, from the development of the polio vaccine to HIV treatment. Yet while the establishment of the HeLa cell line boosted Gey's scientific career, neither Henrietta, who died in 1951, nor her descendants knew anything about the cells until 1973, when researchers studying HeLa cells at Johns Hopkins Hospital approached Henrietta's children for blood samples. The wider public, meanwhile, knew almost none of this until Rebecca Skloot's bestselling book, *The Immortal Life of Henrietta Lacks,* told Henrietta's story in 2010.[254]

More recently, in 2019, the Wellcome Sanger Institute in the UK was accused by whistleblowers of making plans to commercialize a genetic testing product without the consent of hundreds of African people whose DNA samples were used in its development. Gurdasani later identified herself as one of the whistleblowers in a social media post, in which she wrote that she and another whistleblower had lost their jobs at the institute after they had raised concerns to senior management (the Wellcome Sanger Institute declined to comment on personnel when asked about this afterward by *Science* magazine). *The Times* reported at the time that the samples had been collected by scientists at the Lebanese American University and at several African universities, including Stellenbosch University in South Africa, which complained to the Wellcome Sanger Institute. Some of the samples in question came from Indigenous peoples, such as the Nama people from South Africa, Namibia, and Botswana, who had been explicitly informed that their samples would only be used to study population history and human evolution. The Wellcome Sanger Institute said it didn't commercialize or profit from the product—a gene chip—but it later acknowledged in a statement to *Science* magazine that its relationship with some of its research collaborators in Africa had been "disrupted."[255]

Separately, the rise of DNA sequencing also resulted in fresh salt being rubbed into the wounds of the Lacks family. In 2013, the European Molecular Biology Laboratory in Heidelberg, Germany, sequenced and published the HeLa genome without their permission. At the time, Henrietta's descendants argued that the published genome could potentially reveal genetic traits of family members, compromising their privacy. Eventually, the family and the researchers reached a compromise: under the HeLa Genome Data Use Agreement, two members of the Lacks family sit on a US National Institutes of Health working group that grants permission to access HeLa sequence information.[256]

Although the histories of European slavery and colonialism each carry countless examples of exploitation of marginalized people by Western medical institutions and authorities, these problems aren't unique to Western medicine. In 2017, Chinese authorities in Xinjiang came under criticism from human rights activists after it emerged that they were conducting a compulsory scheme of biodata collection, including DNA and blood types, from millions of residents belonging to the Uyghur ethnic minority group. China's Communist Party launched a campaign in 2016 to convert Uyghur people and people belonging to other mainly Muslim ethnic minority groups into loyal supporters, and there are concerns that biodata is being used as yet another tool for surveillance and oppression of these groups. As of the time of writing, authorities are thought to have forcibly detained more than one million Uyghur people in Xinjiang in so-called "reeducation camps," and several countries have accused China of committing genocide against Uyghur people in the region (China has denied these allegations).[257]

The NGO Human Rights Watch reported at the time that DNA and blood type data had been collected through a free program of annual physical exams called Physicals for All, but that it wasn't clear whether participants had been informed of the authorities' intentions to collect, store, or use these sensitive health data. It said that the stated goals of the Physicals for All program included improving the service delivery of health authorities, screening people for diseases, and establishing electronic health records for Xinjiang residents, noting that Chinese media reports had featured testimonies from participants describing how they had received treatments for previously undiagnosed illnesses, in some cases saving their lives. An anonymous Uyghur person who participated in the same program in 2016 told Human Rights Watch that he didn't feel there was an option not to participate, as this "would surely be seen as a sign of 'thought problem,'" meaning political disloyalty. He said the health authorities hadn't provided him with any results following his physical exam.[258]

The program was ongoing as of 2019, when the *New York Times* reported that US company Thermo Fisher Scientific, which had been criticized for supplying DNA sequencing equipment to Xinjiang authorities, had announced it would no longer sell its equipment in the region. In a statement to the newspaper, the Xinjiang government denied collecting DNA samples as part of free medical checkups and said that the DNA sequencing equipment purchased by authorities there was for internal use.[259]

Ethical concerns related to medical research aren't limited to DNA sequencing. In 1996, the city of Kano in northern Nigeria experienced the worst outbreak of meningitis in its history, with more than 120,000 people infected. Several NGOs, including Médecins Sans Frontières (Doctors Without Borders), sent teams to a local hospital to provide support. At the same time the pharmaceutical giant Pfizer, which was in the midst of testing a meningitis drug, also sent a team over. Babatunde Irukera, a lawyer who would later represent Nigeria in a lawsuit filed against the drug company, told the *Financial Times* in 2020 that Pfizer had seen an opportunity for a phase III clinical trial and "scrambled together a team" to conduct it. Irukera noted that the Pfizer staff hadn't worn any badges to clearly distinguish themselves from the humanitarian workers already there. "People were sending their children to them thinking they were going to treat them," he told the newspaper. In 2009, Pfizer settled the resulting lawsuit for $75 million. In a statement announcing the settlement, it denied any wrongdoing or liability, stating that "the 1996 study was conducted with the approval of the Nigerian government and the consent of the participants' parents or guardians, and was consistent with Nigerian laws." Irukera commented to the *Financial Times*: "Pfizer claims it conducted informed consent, but we have no evidence of that. These are people who don't speak English."[260]

Lack of trust is so often cited as a factor underlying racial and ethnic disparities in health, in a way that to me often smells of judgment toward those people from marginalized groups who are wary of the medical establishment. But considering the numerous examples of trust violations by doctors and medical researchers, I think such skepticism is extremely understandable. Trust is easily broken but difficult to build—or *rebuild*.

In 1997, President Bill Clinton issued an official apology on behalf of the US government for a syphilis study carried out at the Tuskegee Institute in Alabama in which researchers deliberately withheld treatment from 399 Black men with the disease. Originally called the "Tuskegee Study of Untreated Syphilis in the Negro Male" (it is now referred to as the US Public Health Services Syphilis Study at Tuskegee), the study involved 600 Black men, including 399 with syphilis and 201 who didn't have the disease. Participants were given free medical exams, free meals, and burial insurance in exchange for their participation, but informed consent wasn't collected, and researchers told the men that they were being treated for "bad blood"—a term used locally at the time to refer to a number of ailments, including syphilis. Even though the antibiotic penicillin had become widely available as a treatment for syphilis and other bacterial diseases by 1943, participants in the study still weren't offered treatment. The study was only stopped almost three decades later in 1972, after an investigation by the Associated Press exposed the injustice. A class-action lawsuit filed on behalf of the study participants and their families the following year resulted in a $10 million out-of-court settlement. One hundred twenty-eight of the study participants died from syphilis or related complications, while forty of their partners were infected and nineteen children were born with congenital syphilis as a consequence of the study.[261]

In his 1997 apology speech at the White House, attended by eight survivors of the study, President Clinton said the study had "served to sow distrust of our medical institutions, especially where research is involved." In fact, the syphilis study at Tuskegee represents just one example of what Latifa Jackson, a geneticist at Howard University in Washington, DC, described to me as a "checkered history of interactions with exploitative governmental research institutions" for Black people and people belonging to other marginalized groups in the US. "I think that the foundation of the medical industry is built on the exploitation of African American [people]," she told me. "No one can even approach biomedical science diagnostics and treatments without using HeLa cells, the unconsented cancer cells collected from Henrietta Lacks, an African American woman," Jackson noted. She is among those who worry that present-day efforts by researchers at academic institutions or in the biomedical industry to increase diversity in their studies represent a repackaged form of this past exploitation. "Genetic discoveries extract resources from underrepresented communities while contributing nothing back to those communities in the form of genomic knowledge, royalties or educational opportunities," she said. "In order to have diverse community participation [in medical research], you need organizational leaders who can establish trust with underrepresented communities," said Jackson.

In Africa, where researchers are increasingly interested in collecting samples from the continent's diverse populations, trust in medical research—and particularly "helicopter science," as 54gene cofounder Abasi Ene-Obong termed it when we spoke—has also been eroded by the numerous instances in which it has been broken. Deepti Gurdasani feels that genomics researchers should be especially conscious of the ethical implications associated with sequencing people's genomes, particularly in low-resource settings. Genome sequencing can reveal genetic mutations underlying health conditions that study participants might not have known they carried. It is therefore vital that researchers build medical infrastructure when working in regions with limited health care services, Gurdasani told me during one of our conversations. During their study in Uganda, she and her team, which included scientists at the UK Medical Research Council's Uganda Medical Informatics Centre, worked to establish infrastructure so that people participating in the study could access treatment and genetic counseling. "You can't diagnose people with disease and then not have a pathway of care," she said.

Zané Lombard, who was involved in the H3Africa study that in 2020 revealed millions of new gene variants by sequencing diverse African populations, thinks that community engagement should also be considered an essential part of medical research. "That's a very important pillar and cornerstone of the H3Africa consortium," she told me, when we spoke. "One of the main components that is absolutely mandatory at the start of every project is that every project has to have a community engagement plan," she explained. "The studies have a community engagement board, which is made up of important people in the community, who then receive some of that research back and then distribute it to the community," she added. Although some of these engagement efforts were hampered by Covid-19 restrictions, Lombard and her colleagues continued to share their findings with members of the public through social and traditional media. "Overall, the response has been really positive," she said. "There's been some really thoughtful questions from participants," she noted. "I think people are really interested in learning more and understanding more about this kind of research."

Questions raised by participants are also used to make adjustments to studies in some cases, said Lombard, giving an example from another one of her ongoing studies that is focused on rare neurodevelopmental disorders in children. Many of the families involved in that study said that they would find it beneficial to be connected with other families with children experiencing similar rare disorders. "The parents have a real need, for instance, for support groups, and bringing parents together who also have a child who is affected by a developmental disorder," Lombard told me. "And that's something now that we're trying to make sure is implemented in the community," she said.

The collaboration between Maria Chahrour and Leah Seyoum-Tesfa, who together established the study of East African children with autism in Texas, presents another example of how researchers can engage with communities. "I would describe it as a mutual learning journey between the two of us," Chahrour said of her collaboration with Seyoum-Tesfa, when we spoke. "I would say it took probably about two years from when we first met, to actually set up everything and have all these discussions back and forth with the community members, and to incorporate their feedback [into the study design]," Chahrour told me. "And that was all before they started actively enrolling participants in the study," she explained. "There was a ton of listening to do at the beginning," she said. "We did spend a lot of time on that," she emphasized. "It has to be a real partnership between the scientists and the community."

Keolu Fox remains wary and thinks medical research as an institution needs to go a step further. He warns that in some cases, community engagement efforts are insufficient and form part of the "illusion of inclusion," which has become something of a catchphrase for him and inspired not only the title of this chapter but also that of an opinion article he authored in the *New England Journal of Medicine* a few months after our conversation in 2020. Fox argues that genomics studies should be led by people from the groups being studied and must benefit them directly. After all, what good does it do if people's DNA samples contribute to the development of new medications that they won't be able to afford?[262]

Fox, who is Native Hawaiian, is particularly passionate about supporting Indigenous genomics researchers and empowering Indigenous populations to protect their genomic data from what he views as commercial exploitation. In his 2020 *New England Journal of Medicine* article, titled "The Illusion of Inclusion—The 'All of Us' Research Program and Indigenous Peoples' DNA," he raised concerns about the National Institutes of Health (NIH) All of Us research program, which aims to sequence DNA from traditionally underrepresented groups in the US, including Indigenous peoples. He noted that previous government-backed, large-scale human genome sequencing efforts, such as the Human Genome Diversity Project, the International HapMap Project, and the 1000 Genomes Project, "provide examples of the ways in which open-source data have been commodified in the past" and argued that the All of Us program should have "built-in mechanisms to protect against the commodification of Indigenous peoples' DNA." Indeed, he pointed out, data from those other, aforementioned large-scale genome sequencing projects "ultimately enabled the generation of nearly a billion dollars' worth of profits by pharmaceutical and ancestry-testing companies." Although participants in genomics research may benefit indirectly from their participation through the development of new drugs by companies, Fox argued that it isn't clear that such drugs will lead to direct benefits for Indigenous communities—or for other marginalized communities whose DNA may have disproportionately contributed to drug development—for instance, in the form of subsidized medications, royalties, or intellectual property rights.

Remember PCSK9 inhibitors—the cholesterol-lowering drugs that were developed based on gene variants discovered in people with recent African ancestry in the US? Well, those drugs are expensive, and one study has already suggested that due to a complicated reimbursement system, patients who are Black or Hispanic are more likely than White patients to have their prescription of the drugs rejected by their insurance provider (Black and Hispanic people—in addition to American Indian or Alaska Native and Native Hawaiian and other Pacific Islander people—are also more likely than their White counterparts to be uninsured in the first place). Even among patients for whom insurance companies approve their prescriptions, the study found that many are forced to abandon them because they cannot afford the expensive co-pays (out-of-pocket costs).[263]

"Do you actually think that including Indigenous people in genome sequencing studies is going to reduce the health disparities that exist in their communities, or do you think that that information—because it's publicly available and there are no restrictions on data access—is going to get commodified by big pharma?" Fox asked me, during our conversation. He is extremely direct and outspoken—and he doesn't shy away from talking about the difficult ethical issues associated with conducting medical studies, particularly within his field. His thinking on these issues is well-informed, not only by his many years of research experience, but also by his own lived experiences and experiences engaging with Indigenous communities in the US. He told me that early on during his career, he had an encounter while giving a talk that got him thinking more deeply about whether and how people should benefit from participating in research.

The talk was at the National Congress of American Indians. "The president of the Navajo Nation is in the room," Fox said, setting the scene for me. Fox was speaking about genomics research, including its potential benefits for advancing personalized medicine. "This is when I used to sell the snake oil myself," he admitted. "I said, 'This new version of predictive and preventative medicine is going to reduce the widening gap in health disparities.' That used to be my line, I used to say that all the time, I used to use it in grants. And I remember saying it—and I was probably like twenty-eight at the time, I think it was 2014," Fox told me. "This is when it hit me like a ton of bricks," he continued. "This older gentleman in the back raises his hand and he says, 'Why should I give a fuck about genome sequencing technologies when I haven't had clean water on my reservation since 1912?'" Fox recalled. "So, you start to realize, we have some serious infrastructure issues in America like [millions of] people who don't have health care coverage. Meanwhile the NIH is trying to sequence a million people's genomes," he said, referring to the "All of Us" program.

Fox has also thought extensively about how things could be improved. "One way to facilitate a paradigm shift toward equitable benefit sharing would be to ensure that Indigenous people have control of data from Indigenous populations, including digital sequence information," he wrote in his *New England Journal of Medicine* piece. "It will take equitable innovation in this area to ensure that the benefits truly reach 'all of us,'" he concluded.

The kind of equitable innovation Fox was referring to is already taking place. In 2022, I spoke with Krystal Tsosie, an Indigenous geneticist and bioethicist at Arizona State University who is a member and citizen of the Navajo Nation. Tsosie is also cofounder and ethics and policy director at the Native BioData Consortium, a nonprofit research institute and biological data repository for US tribes. "The reason why we started was a simple premise," she told me, when I asked her about the motivation and goals behind the consortium. "Data from tribal community members should ultimately be overseen and governed by tribal community members," she said. Her position is shared by Fox, who is a member of the consortium's board. "I am optimistic, because I believe in self-governing, I believe in the democratization of tools, I believe in technological independence, and I believe in the ingenuity of our people," he told me, when we spoke. "I believe in us and because I believe in us, I believe in our future," he added.

Founded in 2018, the Native BioData Consortium is the first non-profit research institute and biological data repository in the US led by Indigenous scientists and tribal members, as well as the first to be housed within the borders of a Native American reservation—it is located within the Cheyenne River Sioux Tribal reservation in Eagle Butte, South Dakota. The consortium is working to build datasets that can be used for research that benefits tribes directly, particularly when it comes to health. For instance, it has been analyzing gene variants associated with rheumatoid arthritis among American Indians from the Northern Plains, who are disproportionately impacted by this autoimmune condition. But unlike many genomics research initiatives, the focus of the Native BioData Consortium isn't only on gathering *biological* data.[264]

"Our goal is to create a repository of genetic information and biological information, and other types of information—including structural and sociocultural factors, environmental factors—for use of tribal community members," Tsosie told me. This is a different approach to that commonly taken in much population genetics and genomics research, she noted, where researchers often look to DNA alone to try to explain health inequities that exist between populations. "What happens is that they're not even really looking beyond the genome," Tsosie commented. "They're looking at just primarily GWAS data, genome wide association data, exome data, sequencing data, whole genome sequencing data, whatever. And whatever information they can get, like clinical phenotyping data from electronic health records. That's it," she said. This ignores a lot of other real factors related to health, she pointed out—not least, vast structural inequities driven by racism. And it is all the more reason why the Native BioData Consortium is additionally working to train a new generation of geneticists and data scientists on how to conduct what Tsosie calls community-centered research.

"If you were to do a community-centered approach in collecting this type of information, you would actually forge partnerships with communities, and then gain the trust of communities so that they would provide that information willingly, so that you could include and make a more holistic, globally representative model that would actually speak to and attest to these types of structural and cultural factors related to disease," Tsosie explained. "Sometimes that qualitative research is considered logistically burdensome because it takes a long time to establish trust with a community, to co-develop a protocol with a community, to have that community approve that protocol, then to do it—to go out into that community and survey," she said. "That's actually a skill set that a lot of population geneticists don't have," she noted.

Tsosie suspects this is in part due to the rise of data science and bioinformatics approaches in genomics, where researchers are increasingly analyzing secondary genomic data from publicly accessible DNA databases rather than collecting primary data directly themselves. "They don't step beyond their computer terminal," she told me. The Native BioData Consortium is working to address this by providing culturally inclusive training opportunities for the next generation of scientists. In 2021, it started IndigiData—an annual summer workshop for undergraduate and graduate students, delivered by Indigenous data scientists and academics. Its curricula are grounded by the principle of tribal sovereignty—the authority of Indigenous tribes to govern themselves within the US—with particular emphasis on Indigenous data sovereignty and empowerment.

The approach to medical research being championed by the Native BioData Consortium comes in sharp contrast to what has been a long history of researchers taking biological samples from Indigenous populations without proper informed consent and, in some cases, even contributing to stigmatization of participating groups or failing to respect group customs surrounding the dead. Moreover, unlike genomic data collected through government funded research, which are typically deposited in large, open-access repositories such as the database of Genotypes and Phenotypes (dbGAP) in the US, the consortium's database isn't publicly accessible. "The problem with depositing either aggregated or anonymized data into this repository [dbGAP] is that there's no community control or community input on the type of research that researchers can do with the data," Tsosie explained to me. "So, there's nothing to stop a researcher who has absolutely no idea about the unique genetic and cultural histories of each Indigenous group from accessing the data and publishing something that's particularly harmful or stigmatizing, or [that] reifies dangerous biological notions of race, without the community's consent."[265]

When we spoke, Tsosie was keen to emphasize that the Native BioData Consortium doesn't exist to stand in the way of scientific research or progress. "It's not blocking research by any means," she told me. "It's just ensuring that the community has consent and oversight over their data," she explained. "If we're going to talk about equalizing power dynamics related to research, then, obviously, that is one step toward justice." Tsosie also believes that the consortium's approach will enhance the quality of the research that gets carried out. "When community members are involved in research, ultimately that enriches research," she said. "Indigenous peoples, who have lived in their land since time immemorial, have seen the changes of environment, changes to their lifestyle, changes to their ways of living, and have seen how that directly impacts their health status," she pointed out. "So, they probably know more about all of the structural barriers to health and cultural factors and colonial factors that drive health inequities in our communities better than an outsider researcher."

The consortium's database has already attracted interest from pharmaceutical companies, who want to use it as a resource to inform drug development, Tsosie told me. "What they were interested in was in utilizing, potentially, access to Indigenous datasets and Indigenous genomes to create a drug that was marketable to the mass population," she explained. But, when she and her colleagues asked, the companies couldn't explain how this would directly benefit those people who had contributed their data. Perhaps the data could be used to inform the development of treatments for health conditions known to disproportionately affect Indigenous populations, Tsosie and her colleagues suggested. The responses from the companies weren't encouraging. "There's no money in it, because it's a small population," Tsosie recalled being told. "It's commercial exploitation. It's bioprospecting, biocommercialism," she sighed. "You're utilizing an untested, novel population—Indigenous peoples—to create a commercial product that isn't ultimately a benefit to them but is a benefit to the company's commercial bottom line," she said.

The question of how and to what extent research participants should benefit from the research they take part in isn't a straightforward one. Some people might be okay with sharing their DNA sequence information for a specific and well-defined purpose, such as for researching a disease that affects them or one of their loved ones directly. They might be happy to do so for free, in the knowledge that it could benefit them, their loved ones, or other people who experience the same disease in the future. Other people might expect more immediate and tangible benefits for contributing their DNA, such as financial compensation. This is pretty much the model of the US-based public benefit corporation LunaDNA; people can get shares of the company in exchange for sharing their genetic data and other information related to their health, such as their diet and exercise habits. Some people may consider it unethical to pay people for their biological data. Perhaps some of those who object to a model of financial compensation would themselves be perfectly happy to donate their DNA for free to any research endeavor that benefits society at large. But even in such a case, people's definitions of societal benefit will vary widely. What if you give away your genetic information for free and then it is used by a private company to develop a drug, from which the company later profits massively? What if that company sells its drug at an extremely high price, such that it is inaccessible to most of the world's population? And what if that same company had previously taken and analyzed samples from you or members of your community without obtaining informed consent? Would you still be willing to give away your data for free?[266]

Ultimately, I think these questions should be for the individuals and communities participating in research to answer, not for medical researchers to answer on their behalf. At the same time, I believe all medical researchers have a responsibility to engage critically with potential ethical issues that could arise from their research, particularly when that research involves collecting or analyzing data from people or communities who have long been marginalized. This includes acknowledging, engaging with, and understanding the history of medicine and medical research, and how this history has contributed to prevalent mistrust of the medical establishment among groups of people who have been and continue to be discriminated against.

Conclusion

If there is a single message that I want this book to convey, it is one of hope: that the health gaps we have been examining throughout these pages aren't inevitable. Recognizing that it is *racism*, rather than race, that is the most significant contributor to racial and ethnic health inequities globally means we are acknowledging that change is possible.

We began in the first chapter of this book by learning about Serena Williams and her frightening ordeal following the birth of her daughter in 2017. Her story captured the world's attention and put a spotlight on the vast racial and ethnic disparities in maternal health that exist in the US and globally. Five years later, in an article in *Elle* magazine, Serena reflected on her experience and on those persisting health inequities. While she emphasized that Black women in the US are almost three times as likely to die during or after childbirth than their White counterparts, she also pointed out that these deaths aren't inevitable. "Many of these deaths are considered by experts to be preventable. Being heard and appropriately treated was the difference between life or death for me; I know those statistics would be different if the medical establishment listened to every Black woman's experience," Serena said.[267]

Listening is vital, especially when it comes to health. And this is not only a matter of doctors listening to patients. It is about all of us listening to each other, including listening to those people within our societies who have for too long been marginalized. This starts way before the onset of ill health or the point of seeking health care. It starts even before preventative measures, such as vaccinations or cancer screenings. It starts within our communities with listening to and learning from each other's diverse lived experiences.

Throughout this book, we have heard from people who have experienced firsthand how harmful racism can be to health, including diverse community members, campaigners, patients, medical students, doctors, academic researchers, innovators, policymakers, and more. Now it is time to listen to and learn from these and many other people not mentioned in this book who are collectively working toward an anti-racist future for our health.

We might start where Serena Williams left off, by listening to her and to the many other Black women and birthing people whom she used her platform to speak up for. In the US, plenty of organizations have been actively working for years to advance maternal health equity and reproductive justice, for example. The Black Mamas Matter Alliance, a network of many such organizations, was established following a collaboration between the Center for Reproductive Rights and the SisterSong Women of Color Reproductive Justice Collective in 2013. The organizations published a joint report for the UN Committee on the Elimination of Racial Discrimination the following year, in which they pointed toward actions that the US government could take to tackle maternal mortality and reproductive injustice. These included increasing both general and pregnancy-related health coverage of uninsured women (women of color are disproportionately uninsured in the US); improving access to contraceptive services and maternal health care; training health care providers to avoid racial stereotypes and to provide high-quality care to all women; and providing more social support for recent parents, including paid parental leave. The point is that neither the problem of health inequity nor these proposed solutions are anything new—this information has been out there for years. While there has certainly been progress in terms of research and awareness, policymakers in the US and globally need to listen better and act faster. We need to support the people and community organizations already working to drive change.[268]

In the UK, where, as in the US, Black women are more likely than White women to die during pregnancy or childbirth, the grassroots organization Five X More asked Black women what actions they wanted to see taken to address the disparity. Based on answers submitted via the organization's landmark Black Maternity Experiences Survey, in 2022 Five X More issued six recommendations for the UK government and health care providers. Fittingly, almost all of the recommendations are grounded in better listening to Black women and birthing people. For instance, the organization called for an official annual maternity survey targeted specifically at Black women, as well as for better mechanisms through which Black women can submit feedback or complaints regarding their maternity care. It also recommended increasing funding for community-based efforts and research aimed at improving Black maternal outcomes. A case in point: Five X More, which offers training to health care professionals as well as providing free information and resources to Black women and birthing people, relies heavily on charitable donations. "We've barely been able to pay ourselves," said Tinuke Awe, one of the organization's cofounders, at an event marking the release of its Black Maternity Experiences report in 2022.

As we have been discovering throughout this book, pregnancy and childbirth are quite literally just the beginning of the problem when it comes to racial and ethnic health disparities; governments globally should also invest more in addressing the many other health inequities that exist across all stages of life. In addition to implementing universal health care in countries such as the US where this isn't already the norm, a large portion of that investment should go toward tackling the wider societal inequities that are the major drivers of racial and ethnic health gaps. Again, there is no shortage of ideas on how to address these problems. As an example: a body of research suggests policies such as universal basic income—unconditional income granted to all citizens by the government—could reduce health inequities. Studies of universal basic income policies trialed in Finland and in parts of the US and Canada have indicated both mental and physical health benefits.

Policy interventions have the potential to address widespread inequities in wealth too, which as we saw in chapter 3 is also a significant contributor to health disparities. To address racial inequities in wealth—and health—the US-based economist Darrick Hamilton, whom we heard from earlier, has proposed and advocated for the widespread introduction of publicly funded child trust accounts known as "baby bonds." These would equip every child, from birth, with the means to attain financial security by having wealth once they become adults. In 2021, Connecticut became the first US state to adopt a Hamilton-inspired "baby bonds" scheme, for children eligible for Medicaid. "It's gaining traction, despite the social context of where we live," Hamilton told me.[269]

It is also a no-brainer that investing in protecting our environment, including tackling the climate crisis, will improve everyone's health—and that goes hand-in-hand with stamping out racism and inequity. To achieve these goals, it will be crucial to listen to the many, often Black and Brown people who are bearing the brunt of pollution and climate change globally. This includes people such as Rosamund Kissi-Debrah, who is continuing to lobby the UK government to make clean air a priority. Yet Rosamund told me that at times she feels excluded from environmental campaigning, because of her race. "There's a reason, by the way, Layal, that they say the environment is all White," she said during one of our discussions, referring to the wider environmental movement. "I'm still an outsider," she said. Rosamund isn't alone in feeling this way. In 2020, Ugandan climate activist Vanessa Nakate drew attention to the issue of erasure of environmental campaigners of color after the Associated Press cropped her out of a photo with White climate activists, including Greta Thunberg, Luisa Neubauer, Isabelle Axelsson, and Loukina Tille. (Following backlash, the Associated Press apologized and replaced the cropped photo.) As Rosamund put it, "the environment should be for everybody." She is right, and neither she nor anyone else should be made to feel like an outsider in an area in which they so evidently have expertise and lived experience.[270]

Yoshira Ornelas Van Horne, whom we heard from earlier, is also an environmental health expert by experience, having grown up in a community in Phoenix, Arizona, that is among those bearing the brunt of environmental inequality. She told me she would like to see increased diversity among researchers in her field who conduct studies in neighborhoods and within communities disproportionately affected by environmental harm. "I want to see people with my background be reflected in these spaces, because a lot of the time it's people that haven't even been to these neighborhoods talking about these neighborhoods without even seeing what they look like," she said. "For me, it's not only documenting these disparities, but also hopefully working alongside these community organizations and these community members to actually bring the interventions and make changes."

Rosamund is among those making changes within her community and beyond. She has been one of the leading voices in the campaign to enshrine the right to breathe clean air into UK law. At the time of writing, almost ten years after nine-year-old Ella died due to severe asthma contributed to by air pollution, the UK looks poised to pass the Clean Air (Human Rights) Bill or Ella's Law, which would establish the right to breathe clean air as a basic human right. Among other things, the law would require local authorities to bring air quality up to minimum WHO standards within five years. Breathing cleaner air will be beneficial for everyone's health, but this is also a matter of racial justice; people of color in England and globally are more likely to live in a very highly polluted area than are White people.[271]

Alongside enacting policies aimed at tackling systemic inequities on a national and even global level, we must also reckon with the role that we all play as individuals in perpetuating racism and its health harms within our societies. We have explored how racism within interpersonal interactions—whether in day-to-day life or at the doctor's office—can contribute to racism-related stress and trauma, with all their associated mental and physical health harms. In some cases, such as with colorism, these interactions may even take place within families. We all have a responsibility to challenge racism in our families, schools, and workplaces, as well as at a more systemic level within our societies.

Throughout this book we have discovered how racism not only makes us unwell but can also influence the health care we receive when we become ill, exacerbating existing racial and ethnic inequities. Just as we have seen within society as a whole, both systemic and interpersonal forms of racism exist in medicine and health care. There has been significant reckoning with this issue within the medical profession in recent years, and that work needs to continue. In chapter 7, we saw how, one by one, different fields of medicine—from neuropsychology to nephrology—have been questioning the validity of using race and ethnicity to interpret people's medical tests or even to guide treatment decisions. Yet despite mounting evidence of its harms, race-based medicine hasn't disappeared; medical and scientific institutions globally have a responsibility to change that. Persisting race-based guidelines or recommendations should be reviewed in a systematic (and scientific) way. Meanwhile, regulations should ensure that medical devices and algorithms aren't designed in ways that exacerbate existing racial and ethnic health inequities.

Crucially, all of us—and particularly those of us in the medical profession—must move away from pathologizing race and instead recognize racism as the disease. Delan Devakumar, a doctor and global health researcher at University College London whom we heard from briefly earlier, feels passionately about this. I spoke with him when he was in the midst of coordinating a special series of articles for *The Lancet* demonstrating how detrimental racism and xenophobia are to health. He told me he hoped the series would encourage the medical community in particular to reckon more deeply with these issues (the series, which I touched on in the introduction to this book, was published toward the end of 2022). "I want the standard GP, pediatric doctor, surgeon, all of those people, when they see a patient, to think about socioeconomic environment, but also racism. How is that affecting the patients they see?" he said.

As a global health researcher, and a global citizen, Devakumar is acutely aware of the worldwide nature of the problem. "This happens everywhere, in every society," he told me. "I was born in Sri Lanka, and then came here when I was very young—to the UK that is," he said. "Living in the UK, overt acts of racism certainly were common when I was a child," Devakumar recalled. But he told me he first became aware of racism in Sri Lanka. "Sri Lanka went through a civil war—a twenty-six-year civil war—between the Tamils who were Hindus mostly and the Sinhalese, who were Buddhists. And this is a religious ethnic conflict that killed a hundred thousand people maybe—we don't know the exact number—but that's based on racist ideas. It comes from the same source," he explained. For him, racism isn't a problem specific to any country or people, rather it's a global health issue. That means it is a problem we can all engage with and collectively tackle, both within and beyond the medical establishment.

Alongside his research, Devakumar also hosts an accessible podcast about racism and health. The podcast is just one of many outputs of a wider collective of academics, artists, activists, policymakers, grassroots organizations, and individuals called Race & Health, spearheaded by Devakumar and his colleagues. I am mentioning this to make the point that there is plenty of freely available information out there, which we can all learn from and (I hope) ultimately use to advocate for fairer, healthier societies. To give one more example: Joel Bervell, a Ghanaian American medical student in the US, has built up a following of hundreds of thousands of people across various social media platforms by sharing short, educational videos about racial health disparities and biases in medicine. (Bervell has also shared videos explaining how the signs and symptoms of various health conditions appear on darker skin tones, which on at least one occasion helped someone to identify a precancerous skin lesion before it progressed to melanoma.)[272]

In the age of information, there is no excuse for anyone—from senior health leaders to policymakers—to be ignorant or uninformed. Having said that, I recognize just how difficult it can be to filter through the huge amounts of information we are all bombarded with on a daily basis. I hope this book can help people to make sense of the latest information and research documenting the many damaging health effects of racism. As we saw in chapter 8, though, there is still a lot of information that we are missing. There is a strong case for countries that aren't already doing so to begin ethically collecting race- and ethnicity-disaggregated health data, as this will make it easier to measure global progress toward achieving health equity. We can't fix problems if we don't know they exist. At the same time, as we usher in a new era of clinical and genomics research that is determined to represent all of humanity, we must be wary of reifying unscientific notions of biological differences between racial and ethnic groups. Importantly, medical researchers and innovators must also reckon with the ethical issues associated with their work and ensure that attempts to increase diversity in medical studies are grounded in both inclusivity and equity.

As a former medical researcher myself, I believe we in the medical research community can learn a lot from our colleagues in the social sciences. "If you're looking at social science and history, the work on racism is massive and it's been going for much longer," Devakumar pointed out. I certainly think there has been an overemphasis on searching for genetic explanations for long-standing racial and ethnic health gaps as opposed to engaging in collaborative, community-based research efforts aimed at tackling the social inequities underlying them.

To eliminate racial and ethnic gaps in health, it is crucial that we recognize racism as the urgent, global public health crisis that it is. Tackling this problem is in all our best interests; eliminating inequities in society and within health care will improve global health, increasing our resilience to disease epidemics. Meanwhile, engaging a more diverse range of people in medical research will enhance scientific discovery and—if that research is conducted ethically and equitably—could lead to better medicines and health care for everyone.

There is an enormous amount we can all do to help tackle the problem. We can listen. We can educate ourselves. We can raise awareness.

We can support and uplift the people and community organizations globally, who are already working to drive change. We can be curious, ask questions, and challenge the status quo. We can be data-driven, not only by collecting data but by using those data to inform positive actions. We can treat racism the same way we treat other things that are widely known to be bad for our health. And we can welcome anti-racism as the next major advance in human health.

Acknowledgments

I am extremely thankful to all the people generous enough to share their personal experiences with racism and health publicly, including all those who took the time to speak with me as part of my research for this book. I am also grateful to all those whose prior research, reporting, or general expertise I was able to draw upon, including journalists, authors, academic researchers, health care workers, medical students, patient advocates, campaigners, and activists.

I would also like to thank the many people who helped make this book an actual book, starting with my literary agent, Will Francis. Will understood and supported my vision from day one, and has been my confident and unfaltering guide through the world of publishing ever since. He is brilliant at what he does and is also a lovely person. His colleagues at Janklow & Nesbit are equally wonderful and I am grateful for everything they have done to support this book and me as an author.

Editing is an indispensable yet somewhat invisible process. I am immensely grateful to Alexis Kirschbaum and Jasmine Horsey at Bloomsbury Publishing, and to Alessandra Bastagli and Rola Harb at Astra House, for elevating this book with their thoughtful feedback and editing. I am similarly thankful to Darren Ankrom for thorough fact-checking on behalf of Astra House, and to Ben Brock at Bloomsbury for careful copyediting. I also want to thank Lauren Whybrow at Bloomsbury for seamlessly managing the whole editing process, and everyone else at Janklow & Nesbit, Bloomsbury, and Astra House who contributed toward the successful publication of this book.

This book was heavily inspired by my journalism. Below is a list of chapters that were partly based on articles I authored in *New Scientist* and *Wired*, along with the relevant article information:

CHAPTERS 2-3

Layal Liverpool, "Why Coronavirus Hit People from BAME Communities So Hard," *Wired*, August 2020, https://www.wired.co.uk/article/bame-communities-coronavirus-uk.

CHAPTER 4

Layal Liverpool, "Systemic Racism: What Research Reveals about the Extent of Its Impact," *New Scientist*, November 2020, https://www.newscientist.com/article/mg24833093-900-systemic-racism-what-research-reveals-about-the-extent-of-its-impact/.

CHAPTER 7

Layal Liverpool, "How Medical Tests Have Built In Discrimination against Black People," *New Scientist*, July 2021, https://www.newscientist.com/article/mg25133434-100-how-medical-tests-have-built-in-discrimination-against-black-people/.

CHAPTERS 8-10

Layal Liverpool, "Genomic Medicine Is Deeply Biased towards White People," *New Scientist*, January 2021, https://www.newscientist.com/article/mg24933180-800-genomic-medicine-is-deeply-biased-towards-white-people/.

New Scientist is the place where I learned how to be a science journalist, first as an intern and then as a trainee reporter. I am enormously thankful to Jacob Aron, Timothy Revell, Penny Sarchet, Emily Wilson, and Nina Wright for giving me the opportunity to become part of the *New*

Scientist team in 2019 and for everything they have done to support my career since. I am also hugely grateful to Adam Vaughan, Chelsea Whyte, and Clare Wilson for mentoring me and helping me to find opportunities to expand my journalism at the start of my career. The *New Scientist* articles mentioned above were made infinitely better by editing from Jacob Aron and Daniel Cossins, as well as sub-editing from Eleanor Parsons and her team. I would also like to thank all of my other colleagues at *New Scientist* for their encouragement, mentorship, and peer support, including fellow 2019/2020 interns Gege Li and Jason Arunn Murugesu.

I am additionally thankful to Matt Reynolds, who conceived of and edited the *Wired* article mentioned above, and whose input and guidance was invaluable at an early stage of my career.

I would also like to thank my colleagues and mentors at *Nature*, especially my manager, Nisha Gaind, who was extremely supportive and understanding when I was in the final stages of editing this book and who is an all-around delight to work with. Thanks also to my former mentors at *The Guardian*, Hannah Devlin and Ian Sample.

So many people helped me on the path from vague idea in my head to book proposal to book, and I am enormously grateful to them all. Jenny Lord gave me the push I needed to start seriously thinking about writing a book on this important subject. I am grateful to Adam Rutherford and Angela Saini, both brilliant authors on racism in science—they are both great inspirations to me and their encouragement helped to convince me that writing this book was something I could—and should—do. Adam and Angela, as well as Michael Marshall and Jennifer Pinkowski, helped to demystify for me the path toward becoming an author. I am really grateful for the time they gave me and for their openness and honesty in sharing their own experiences of book writing and publishing.

Other nonfiction authors on the subjects of racism, inequality, and health who have inspired me are Akala, Caroline Criado-Perez, Emma Dabiri, Reni Eddo-Lodge, Alice Hasters, Afua Hirsch, Ibram X. Kendi,

Natalie Morris, Derecka Purnell, Dorothy E. Roberts, Rebecca Skloot, Annabel Sowemimo, Steven Thrasher, Linda Villarosa, and Harriet A. Washington—and that's to name just a few. I have also been inspired by the writing of many fiction authors, particularly Chimamanda Ngozi Adichie, Malorie Blackman, Bernadine Evaristo, Yaa Gyasi, and Zadie Smith.

Thank you to my schoolteachers Adam Bushnell, Mariske Jonker, and Clare van der Willigen, who collectively nurtured my passions for science and writing from an early age.

I am so grateful to Diana Kwon and the rest of the Berlin science journalists group for their valuable peer support, and especially to Diana for her friendship. I am also grateful to be part of the Association of British Science Writers and its German equivalent, the Wissenschaftspressekonferenz, both of which provide excellent resources for writers.

Before I became a writer, I worked as a biomedical researcher in the UK at University College London (UCL) and at the University of Oxford. My experiences working in medical research were extremely influential for this book. I would like to thank some of my science mentors. Thank you to Richard Milne, Greg Towers, and Chris Van Tulleken for cheering me on consistently in my career, starting at UCL and continuing even as I have deviated far from laboratory science. I am grateful to all my other sciencey friends as well, and would like to particularly shout out Richard Morter, who is my go-to source for anything medicine related, including any ailments that happen to afflict me or my family members—as well as Cherrelle Dacon, Antonio Gregorio Dias Jr., and Ugochim Eduputa, who consistently keep me up to speed both with the latest biomedical research, and Beyoncé's latest music. I am also lucky to have many smart, talented friends who aren't biomedical scientists, and I have been similarly grateful for their support, encouragement, and input while I've been working on this book. Thank you, Matita Arize, Anna Balawejder, Minodora Cristescu, Zbigniew Malinowski, Nicole Mancienne, Camilla Omollo, Antje Radünz, and Eline Reintjes. Eline in particular has generously responded to countless WhatsApp messages from me about this book and been a huge cheerleader from the beginning.

I also want to thank my family for their never-ending love, support, and encouragement. Thanks Krzyś, Asia, Eric, and Marieke. Thank you to my father, Philip, for asking me tough questions and challenging me to debates that helped me to hone some of the arguments in the book. Special thanks go to my mother, Rawia, and my sister, Yasmin, both of whom kindly agreed to share their personal stories for this book, and who are part of my constant cheerleading squad in life. Captain of that cheerleading squad is my husband, Konrad, without whose love, patience, encouragement, and support there would be no book—thank you, my sweetheart.

Advocacy Groups and Resources

Below is a list of organizations, advocacy groups, campaigns, and other resources referred to in the book.

The AAKOMA Project: A nonprofit organization focused on illuminating and reducing mental health disparities for diverse communities in the US. https://aakomaproject.org/

The Asian American and Native Hawaiian/Pacific Islander Data Repository (AAPI Data): An organization that publishes demographic data and policy research on Asian Americans, Native Hawaiians, and Pacific Islanders in the US. https://aapidata.com/

Afrozensus: A large-scale online survey of Black, African, and Afro-diasporic people in Germany. https://afrozensus.de/

Agents for Change in Environmental Justice: A program to amplify the voices, stories, podcasts, and research of diverse emerging leaders in environmental justice. https://agentsofchangeinej.org/

Birthing on Country: An international social justice movement to redress the negative impact of colonization and return childbirth services to First Nations communities and First Nations control. https://www.birthingoncountry.com/

Black & Brown Skin: An online platform that showcases images of various health conditions as they appear on darker skin tones. https://www.blackandbrownskin.co.uk/

The Black Mamas Matter Alliance: A Black women–led cross-sectoral alliance of organizations centering Black women and birthing people, and advocating for maternal health equity and reproductive justice. https://blackmamasmatter.org/

The Center for Reproductive Rights: A global human rights organization of lawyers and advocates working to ensure reproductive rights are protected in law as fundamental human rights. https://reproductiverights.org/

Centre for Evidence Based Dermatology Skin of Colour Resource: A resource bringing together information sources, systematic reviews, and review articles on topics of relevance to skin of color. https://www.nottingham.ac.uk/research/groups/cebd/resources/skin-of-colour/index.aspx

The Congress of Aboriginal and Torres Strait Islander Nurses and Midwives (CATSINaM): The collective national voice of Aboriginal and Torres Strait Islander nurses and midwives. https://catsinam.org.au/

The CROWN Act: A law adopted in California and subsequently in several other US states to prohibit race-based hair discrimination, which is the denial of employment and educational opportunities because of hair texture or protective hairstyles including braids, locs, twists, or bantu knots. "CROWN" stands for "Creating a Respectful and Open World for Natural Hair." https://www.thecrownact.com/

Decolonising Contraception: A UK-based charity committed to developing an inclusive approach to reproductive health and rights. https://decolonisingcontraception.com/

The D'lissa Christina Princess Foundation: A UK-based foundation launched on the first anniversary of D'lissa Parkes's death, in her memory, and on her daughter S'riaah Christina Parkes's first birthday. The purpose of the foundation is to raise awareness of the issues of maternal deaths and health, and to provide a support network for women and men of all ages and backgrounds. https://www.dcpfoundation.co.uk/

The Ecological Waste Coalition of the Philippines (EcoWaste Coalition): An advocacy group campaigning for an eco-friendly, zero waste, and toxic-free Philippines. https://www.ecowastecoalition.org/

The Ella Roberta Family Foundation: A UK-based charity campaigning for clean air for all, founded in memory of Ella Roberta Adoo Kissi-Debrah. https://ellaroberta.org/

The European Network of Equality Bodies (Equinet): A membership organization promoting equality in Europe by supporting and enabling the work of national equality bodies. https://equineteurope.org/

The European Network Against Racism (ENAR): A European Union–wide network of anti-racist nongovernmental organizations. https://www.enar-eu.org/

Five X More: A grassroots organization committed to changing Black women and birthing people's maternal health outcomes in the UK. https://fivexmore.org/

The Halo Collective: An alliance of organizations and individuals working to create a future without hair discrimination. https://halocollective.co.uk/

Hutano: A social health platform for Black and Brown people. https://www.tryhutano.com/

Melanin Medics: A nonprofit organization that supports aspiring and current doctors of African and Caribbean heritage in the UK. https://www.melaninmedics.com/

The NHS Race and Health Observatory: An independent expert body funded by NHS England, which works to identify and tackle ethnic inequalities in health and care. https://www.nhsrho.org/

The National Justice Project: An independent human rights legal service based in Australia. https://justice.org.au/

Native Biodata Consortium: A nonprofit research institute and biological data repository for US tribes. https://nativebio.org/

Projeto Inclusão (The Inclusion Project): A grassroots organization working to support low-income families in São João de Meriti, Brazil. https://www.facebook.com/pieden2014 and https://www.instagram.com/inclusaoprojeto/

Race & Health: A collective of academics, artists, activists, policymakers, grassroots organizations, and individuals aiming to reduce the adverse effects of discrimination that lead to poor health. https://www.raceandhealth.org/

Reaching Families Advocacy and Support Group (REACH): A nonprofit organization in the US that supports East African immigrant families with disabled children. https://reachfamilies.org/

Roots Community Birth Center: An African American–owned, midwife-led, freestanding birth center in Minneapolis, Minnesota. https://www.rootsbirth center.com/

The Runnymede Trust: A race-equality think tank in the UK. https://www .runnymedetrust.org/

The Sickle Cell Society: A nonprofit organization in the UK supporting and representing people affected by sickle cell disorder to improve their overall quality of life. https://www.sicklecellsociety.org/

SisterSong Women of Color Reproductive Justice Collective: A national membership organization based in the southern US working to build a network of individuals and organizations to improve institutional policies and systems that impact the reproductive lives of marginalized communities. https://www .sistersong.net/

#SmearYourMea: A community-driven campaign in New Zealand set up by the late Talei Morrison to raise awareness and promote advocacy and support through early detection, treatment, and prevention of cervical cancer in Māori communities.

Visual Atlas: A collection of images established at the University of New Mexico featuring dermatological conditions on a range of skin tones. https://hsc.unm .edu/medicine/departments/dermatology/inclusive-dermatology/

References

1. Cathy Cassata, "This Med Student Wrote the Book on Diagnosing Disease on Darker Skin," Healthline, May 11, 2021, https://www.healthline.com/health-news/this-med-student-wrote-the-book-on-diagnosing-disease-on-darker-skin.
   Sydney Page, "A Medical Student Couldn't Find How Symptoms Look on Darker Skin. He Decided to Publish a Book About It," *Washington Post*, July 22, 2020, https://www.washingtonpost.com/lifestyle/2020/07/22/malone-mukwende-medical-handbook/.
2. Malone Mukwende, Peter Tamony, and Margot Turner, *Mind the Gap: A Handbook of Clinical Signs in Black and Brown Skin* (2020), available at https://www.blackandbrownskin.co.uk/mindthegap.
   Molly Longman, "How Bias in Medical Textbooks Endangers BIPOC," Refinery29, September 1, 2020, https://www.refinery29.com/en-us/2020/08/9991646/medical-racism-represenation-in-medical-textbooks.
   Clare Roth, "Lingering Colonial Legacies: The Study of Skin Is Too White," Deutsche Welle, December 12, 2022, https://www.dw.com/en/dermatology-study-of-skin-too-white/a-60632976.
   Shani Kohli Kurd and Joel M. Gelfand, "The Prevalence of Previously Diagnosed and Undiagnosed Psoriasis in US Adults: Results from NHANES 2003–2004," *Journal of the American Academy of Dermatology* 60, no. 2 (February 2009), https://doi.org/10.1016/j.jaad.2008.09.022.
   Valerie M. Harvey, Joan Paul, and L. Ebony Boulware, "Racial and Ethnic Disparities in Dermatology Office Visits among Insured Patients, 2005–2010," *Journal of Health Disparities Research and Practice* 9, no. 2 (2016), https://digitalscholarship.unlv.edu/jhdrp/vol9/iss2/6/.
3. William J. Hall et al., "Implicit Racial/Ethnic Bias among Health Care Professionals and Its Influence on Health Care Outcomes: A Systematic Review," *American Journal of Public Health* 105, no. 12 (December 2015), https://doi.org/10.2105/ajph.2015.302903.
   Sophie Trawalter, Kelly M. Hoffman, and Adam Waytz, "Racial Bias in Perceptions of Others' Pain," *PLoS One* 7, no. 11 (November 2012), https://doi.org/10.1371/journal.pone.0048546.
   Laura D. Wandner et al., "The Perception of Pain in Others: How Gender, Race, and Age Influence Pain Expectations," *Journal of Pain* 13, no. 3 (March 2012), https://doi.org/10.1016%2Fj.jpain.2011.10.014.
   Vani A. Mathur et al., "Racial Bias in Pain Perception and Response: Experimental Examination of Automatic and Deliberate Processes," *Journal of Pain* 15, no. 5 (May 2014), https://doi.org/10.1016%2Fj.jpain.2014.01.488.
   Adam Waytz, Kelly M. Hoffman, and Sophie Trawalter, "A Superhumanization Bias in Whites' Perceptions of Blacks," *Social Psychological and Personality Science* 6, no. 3 (April 2015), https://doi.org/10.1177/1948550614553642.

Astha Singhal, Yu-Yu Tien, and Renee Y. Hsia, "Racial-Ethnic Disparities in Opioid Prescriptions at Emergency Department Visits for Conditions Commonly Associated with Prescription Drug Abuse," *PLoS One* 11, no. 8 (2016), https://doi.org/10.1371/journal.pone.0159224.

Kelly M. Hoffman et al., "Racial Bias in Pain Assessment and Treatment Recommendations, and False Beliefs about Biological Differences between Blacks and Whites," *Proceedings of the National Academy of Sciences* 113, no. 16 (April 2016), https://doi.org/10.1073/pnas.1516047113.

4. Ana Valdes, "Coronavirus: BAME Deaths Urgently Need to be Understood, Including Any Potential Genetic Component," *The Conversation*, May 21, 2020, https://theconversation.com/coronavirus-bame-deaths-urgently-need-to-be-understood-including-any-potential-genetic-component-138400.

   Nicole Phillips et al., "The Perfect Storm: COVID-19 Health Disparities in US Blacks," *Journal of Racial and Ethnic Health Disparities* 8, no. 5 (October 2021), https://doi.org/10.1007%2Fs40615-020-00871-y.

5. "Medicine and Medical Science: Black Lives Must Matter More," *The Lancet* 395, no. 10240 (June 2020), https://doi.org/10.1016/s0140-6736(20)31353-2.

   Delan Devakumar et al., "Racism, Xenophobia, Discrimination, and the Determination of Health," *The Lancet* 400, no. 10368 (December 2022), https://doi.org/10.1016/s0140-6736(22)01972-9.

   Strengthening Primary Health Care to Tackle Racial Discrimination, Promote Intercultural Services and Reduce Health Inequities, World Health Organization, 2022, https://www.who.int/publications/i/item/9789240057104.

6. "Sen. Bill Cassidy on His State's Racial Disparities in Coronavirus Deaths," NPR, April 7, 2020, https://www.npr.org/2020/04/07/828715984/sen-bill-cassidy-on-his-states-racial-disparites-in-coronavirus-deaths.

   The Report of the Commission on Race and Ethnic Disparities, UK Commission on Race and Ethnic Disparities, 2021, https://www.gov.uk/government/publications/the-report-of-the-commission-on-race-and-ethnic-disparities.

7. Dorothy Roberts, *Fatal Invention: How Science, Politics, and Big Business Re-create Race in the Twenty-First Century* (New York: New Press, 2012).

   Angela Saini, *Superior: The Return of Race Science* (Boston: Beacon Press, 2019).

   Adam Rutherford, *How to Argue with a Racist: History, Science, Race and Reality* (London: Weidenfeld & Nicolson, 2020).

   Annabel Sowemimo, *Divided: Racism, Medicine and Why We Need to Decolonise Healthcare* (Wellcome Collection, 2023).

8. Adam Rutherford, *Control: The Dark History and Troubling Present of Eugenics* (London: Weidenfeld & Nicolson, 2022).

9. Raj Bhopal, "The Beautiful Skull and Blumenbach's Errors: The Birth of the Scientific Concept of Race," *BMJ* 335, no. 7633 (December 2007), https://doi.org/10.1136%2Fbmj.39413.463958.80.

10. Julia Galway-Witham and Chris Stringer, "How Did *Homo sapiens* Evolve?," *Science* (1979) 360, no. 6395 (June 2018), https://doi.org/10.1126/science.aat6659.

11. R. C. Lewontin, "The Apportionment of Human Diversity," in *Evolutionary Biology* (Springer, 1972), https://link.springer.com/chapter/10.1007/978-1-4684-9063-3_14.

12. Grace Huckins, "Richard Lewontin Leaves a Legacy of Fighting Racism in Science," *Wired*, September 3, 2021, https://www.wired.com/story/richard-lewontin-leaves-a-legacy-of-fighting-racism-in-science/.

13. Noah A. Rosenberg et al., "Genetic Structure of Human Populations," *Science* 298, no. 5602 (December 20, 2002), https://doi.org/10.1126/science.1078311.

14. Nicholas Wade, "Gene Study Identifies 5 Main Human Populations, Linking Them to Geography," *New York Times*, December 20, 2002, https://www.nytimes.com/2002/12

/20/us/gene-study-identifies-5-main-human-populations-linking-them-to-geography.html.

Sarah A. Tishkoff and Kenneth K. Kidd, "Implications of Biogeography of Human Populations for 'Race' and Medicine," *Nature Genetics* 36, S21–S27 (2004), https://doi.org/10.1038/ng1438.

Adam Rutherford, *A Brief History of Everyone Who Ever Lived: The Stories in Our Genes* (London: Weidenfeld & Nicolson, 2016).

15. Deborah Posel, "Race as Common Sense: Racial Classification in Twentieth-Century South Africa," *African Studies Review* 44, no. 2 (September 2001), https://doi.org/10.2307/525576.

    Christine B. Hickman, "The Devil and the One Drop Rule: Racial Categories, African Americans, and the US Census," *Michigan Law Review* 95, no. 5 (March 1997), https://repository.law.umich.edu/mlr/vol95/iss5/2/.

16. Rob Haskell, "Serena Williams on Motherhood, Marriage, and Making Her Comeback," *Vogue*, January 10, 2018, https://www.vogue.com/article/serena-williams-vogue-cover-interview-february-2018.

17. Julia Reinstein, "Black Women Are Speaking Out after Serena Williams Revealed She Faced Life-Threatening Birth Complications," BuzzFeed News, January 16, 2018, https://www.buzzfeednews.com/article/juliareinstein/serena-williams-birth-complications.

    Centers for Disease Control and Prevention: Pregnancy Mortality Surveillance System, "Pregnancy-Related Deaths by Race/Ethnicity (1987–2017)."

    Russell Fuller, "Serena Williams: Statistics on Deaths in Pregnancy or Childbirth 'Heart-Breaking,'" BBC, March 6, 2018, https://www.bbc.com/sport/tennis/43299147.

18. Centers for Disease Control and Prevention: Pregnancy Mortality Surveillance System, "About the Pregnancy Mortality Surveillance System (PMSS)," https://www.cdc.gov/reproductivehealth/maternal-mortality/pregnancy-mortality-surveillance-system.htm#about-pmss.

19. Centers for Disease Control and Prevention: National Center for Health Statistics, "Detailed Evaluation of Changes in Data Collection Methods," https://www.cdc.gov/nchs/maternal-mortality/evaluation.htm.

    Donna L. Hoyert, "Maternal Mortality Rates in the United States, 2019," NCHS Health E-Stats, April 2021, https://doi.org/10.15620/cdc:103855.

    Katy B. Kozhimannil et al., "Severe Maternal Morbidity and Mortality among Indigenous Women in the United States," *Obstetrics and Gynecology* 135, no. 2 (February 2020), https://doi.org/10.1097%2FAOG.0000000000003647.

20. Emily E. Petersen et al., "Racial/Ethnic Disparities in Pregnancy-Related Deaths—United States, 2007–2016," *Morbidity and Mortality Weekly Report* 68, no. 35 (September 6, 2019), https://doi.org/10.15585/mmwr.mm6835a3.

    Ray Boshara, William R. Emmons, and Bryan J. Noeth, "The Demographics of Wealth—How Age, Education and Race Separate Thrivers from Strugglers in Today's Economy. Essay No. 2: The Role of Education," *Demographics of Wealth*, Federal Reserve Bank of St. Louis (2015), https://ideas.repec.org/a/fip/fedldw/00002.html.

21. *Nowhere To Go: Maternity Care Deserts Across The US (2020 Report)*, March of Dimes (2020), https://www.marchofdimes.org/maternity-care-deserts-report.

    Peiyin Hung et al., "Access to Obstetric Services in Rural Counties Still Declining, with 9 Percent Losing Services, 2004–14," *Health Affairs* 36, no. 9 (September 2017), https://doi.org/10.1377/hlthaff.2017.0338.

22. Elizabeth A. Howell et al., "Site of Delivery Contribution to Black-White Severe Maternal Morbidity Disparity," *American Journal of Obstetrics and Gynecology* 215, no. 2 (August 2016), https://doi.org/10.1016/j.ajog.2016.05.007.

    Andreea A. Creanga et al., "Performance of Racial and Ethnic Minority-Serving Hospitals on Delivery-Related Indicators," *American Journal of Obstetrics and Gynecology* 211, no. 6 (December 2014), https://doi.org/10.1016/j.ajog.2014.06.006.

Elizabeth A. Howell et al., "Black-White Differences in Severe Maternal Morbidity and Site of Care," *American Journal of Obstetrics and Gynecology* 214, no. 1 (January 2016), https://doi.org/10.1016/j.ajog.2015.08.019.

Ben Horowitz et al., "Systemic Racism Haunts Homeownership Rates in Minnesota," Federal Reserve Bank of Minneapolis, February 25, 2021, https://www.minneapolisfed.org/article/2021/systemic-racism-haunts-homeownership-rates-in-minnesota.

23. *MBRRACE-UK Saving Lives, Improving Mothers' Care—Lessons Learned to Inform Maternity Care from the UK and Ireland Confidential Enquiries into Maternal Deaths and Morbidity 2017–19*, MBRRACE-UK (2021), https://www.npeu.ox.ac.uk/assets/downloads/mbrrace-uk/reports/maternal-report-2021/MBRRACE-UK_Maternal_Report_2021_-_FINAL_-_WEB_VERSION.pdf.

    Black People, Racism and Human Rights, House of Commons House of Lords Joint Committee on Human Rights, 2020, https://committees.parliament.uk/publications/3376/documents/32359/default/.

24. Marian Knight et al., "A National Population-Based Cohort Study to Investigate Inequalities in Maternal Mortality in the United Kingdom, 2009–17," *Paediatric and Perinatal Epidemiology* 34, no. 4 (July 2020), https://doi.org/10.1111/ppe.12640.

    Marian Knight et al., "Characteristics and Outcomes of Pregnant Women Admitted to Hospital with Confirmed SARS-COV-2 Infection in UK: National Population Based Cohort Study," *BMJ* 369 (2020), https://doi.org/10.1136/bmj.m2107.

25. Black Maternal Healthcare and Mortality, vol. 692, Debated on Monday, April 19, 2021, UK Parliament, https://hansard.parliament.uk/Commons/2021-04-19/debates/6935B9C7-6419-4E7B-A813-E852A4EE4F5C/BlackMaternalHealthcareAndMortality.

26. Manisha Nair, Jennifer J. Kurinczuk, and Marian Knight, "Ethnic Variations in Severe Maternal Morbidity in the UK—a Case Control Study," *PLoS One* 9, no. 4 (April 2014), https://doi.org/10.1371/journal.pone.0095086.

27. Suresh Jungari and Bal Govind Chauhan, "Caste, Wealth and Regional Inequalities in Health Status of Women and Children in India," *Contemporary Voice of Dalit* 9, no. 1 (April 2017), https://doi.org/10.1177/2455328X17690644.

    Mukesh Hamal et al., "Social Determinants of Maternal Health: A Scoping Review of Factors Influencing Maternal Mortality and Maternal Health Service Use in India," *Public Health Reviews* 41, no. 13 (2020), http://dx.doi.org/10.1186/s40985-020-00125-6.

    J. Li et al., "Maternal Mortality in Yunnan, China: Recent Trends and Associated Factors," *BJOG* 114, no. 7 (July 2007), https://doi.org/10.1111/j.1471-0528.2007.01362.x.

    Yan Ren et al., "Ethnic Disparity in Maternal and Infant Mortality and its Health-System Determinants in Sichuan Province, China, 2002–14: An Observational Study of Cross-Sectional Data," *Lancet* 390, no. S7 (December 2017), http://dx.doi.org/10.1016/S0140-6736(17)33145-8.

    Maria do Carmo Leal et al., "The Color of Pain: Racial Iniquities in Prenatal Care and Childbirth in Brazil," *Cadernos de Saúde Pública* 33, suppl. 1 (July 2017), https://doi.org/10.1590/0102-311x00078816.

    L. L. Wall, "Dead Mothers and Injured Wives: The Social Context of Maternal Morbidity and Mortality among the Hausa of Northern Nigeria," *Studies in Family Planning* 29, no. 4 (December 1998), https://doi.org/10.2307/172248.

    Catherine Meh et al., "Levels and Determinants of Maternal Mortality in Northern and Southern Nigeria," *BMC Pregnancy and Childbirth* 19, no. 417 (2019), https://doi.org/10.1186/s12884-019-2471-8.

    Jennie B. Gamlin and Sarah J. Hawkes, "Pregnancy and Birth in an Indigenous Huichol Community: From Structural Violence to Structural Policy Responses," *Culture, Health & Sexuality* 17, no. 1 (2015), https://doi.org/10.1080/13691058.2014.950334.

    Fact sheet: Maternal Mortality, World Health Organization, 2023, https://www.who.int/news-room/fact-sheets/detail/maternal-mortality.

28. Sylvia Parkes, "Five X More Blog: D'lissa Parkes," Five X More, 2020, https://www.fivexmore.org/blog/dlissa-parkes.
29. "Why Is Covid Killing People of Colour," BBC, 2021, https://www.imdb.com/title/tt14146576/.
30. Sam Tobin and Patrick Lion, "Young Mum Who Died During C-Section Was Suffering with Such Severe Constipation Her Womb Was Bent Out of Shape," *Daily Mirror*, October 3, 2016, https://www.mirror.co.uk/news/uk-news/young-mum-who-died-during-8969070.
31. W. E. Caldwell and H. C. Moloy, "Anatomical Variations in the Female Pelvis and their Effect in Labor with a Suggested Classification," *American Journal of Obstetrics and Gynecology* 26, no. 4 (October 1933), https://doi.org/10.1016/S0002-9378(33)90194-5.
    Wm. Turner, "The Index of the Pelvic Brim as a Basis of Classification," *Journal of Anatomy and Physiology* 181, no. 1 (1885), https://www.ncbi.nlm.nih.gov/pmc/articles/PMC1288541/.
32. Pamela L. Geller and Miranda K. Stockett, eds., *Feminist Anthropology: Past, Present, and Future* (Philadelphia: University of Pennsylvania Press, 2006).
    A. H. Schultz, "Sex Differences in the Pelves of Primates," *American Journal of Physical Anthropology* 7, no. 3 (September 1949), https://doi.org/10.1002/ajpa.1330070307.
    Lia Betti, "Human Variation in Pelvic Shape and the Effects of Climate and Past Population History," *Anatomical Record* 300, no. 4 (April 2017), https://doi.org/10.1002/ar.23542.
    R. V. Baragi et al., "Differences in Pelvic Floor Area between African American and European American Women," *American Journal of Obstetrics and Gynecology* 187, no. 1 (July 2002), https://doi.org/10.1067/mob.2002.125703.
    T. Wingate Todd and Anna Lindala, "Dimensions of the Body: Whites and American Negroes of Both Sexes," *American Journal of Physical Anthropology* 12, no. 1 (July/September 1928), https://doi.org/10.1002/ajpa.1330120104.
33. Algis Kuliukas et al., "Female Pelvic Shape: Distinct Types or Nebulous Cloud?," *British Journal of Midwifery* 23, no. 7 (July 2015), http://dx.doi.org/10.12968/bjom.2015.23.7.490.
    Lia Betti and Andrea Manica, "Human Variation in the Shape of the Birth Canal Is Significant and Geographically Structured," *Proceedings: Biological Sciences* 285, no. 1889 (October 2018), https://doi.org/10.1098/rspb.2018.1807.
34. Elizabeth O'Brien, "Pelvimetry and the Persistence of Racial Science in Obstetrics," *Endeavour* 37, no. 1 (March 2013), https://doi.org/10.1016/j.endeavour.2012.11.002.
35. Rupa Marya and Raj Patel, *Inflamed: Deep Medicine and the Anatomy of Injustice* (Allen Lane, 2021) 202–208.
    D. Ojanuga, "The Medical Ethics of the 'Father of Gynecology, Dr J Marion Sims," *Journal of Medical Ethics* 19, no. 1 (March 1993), https://doi.org/10.1136%2Fjme.19.1.28.
36. Michelle Peter and Reyss Wheeler, *The Black Maternity Experiences Report*, Five X More, May 2022, https://www.fivexmore.org/blackmereport.
37. *Systemic Racism, Not Broken Bodies: An Inquiry into Racial Injustice and Human Rights in UK Maternity Care*, Birthrights, 2022, https://www.birthrights.org.uk/campaigns-research/racial-injustice/.
38. Jane Henderson, Haiyan Gao, and Maggie Redshaw, "Experiencing Maternity Care: The Care Received and Perceptions of Women from Different Ethnic Groups," *BMC Pregnancy and Childbirth* 13, no. 196 (2013), https://doi.org/10.1186/1471-2393-13-196.
    Julie Jomeen and Maggie Redshaw, "Ethnic Minority Women's Experience of Maternity Services in England," *Ethnicity & Health* 18, no. 3 (2013), https://doi.org/10.1080/13557858.2012.730608.
39. J. W. Collins Jr. et al., "Low-Income African-American Mothers' Perception of Exposure to Racial Discrimination and Infant Birth Weight," *Epidemiology* 11, no. 2 (May 2000), https://doi.org/10.1097/00001648-200005000-00019.

Lynn Rosenberg et al., "Perceptions of Racial Discrimination and the Risk of Preterm Birth," *Epidemiology* 13, no. 6 (November 2002), https://doi.org/10.1097/00001648-200211000-00008.

Tyan Parker Dominguez et al., "Racial Differences in Birth Outcomes: The Role of General, Pregnancy, and Racism Stress," *Health Psychology* 27, no. 2 (March 2008), https://doi.org/10.1037/0278-6133.27.2.194.

Jeff A. Dennis, "Birth Weight and Maternal Age Among American Indian/Alaska Native Mothers: A Test of the Weathering Hypothesis," *SSM Population Health* 7 (April 2019), https://doi.org/10.1016%2Fj.ssmph.2018.10.004.

Bani Saluja and Zenobia Bryant, "How Implicit Bias Contributes to Racial Disparities in Maternal Morbidity and Mortality in the United States," *Journal of Women's Health* 30, no. 2 (February 2021), https://doi.org/10.1089/jwh.2020.8874.

Jamila K. Taylor, "Structural Racism and Maternal Health among Black Women," *Journal of Law, Medicine & Ethics* 48, no. 3 (September 2020), https://doi.org/10.1177/1073110520958875.

Saraswathi Vedam et al., "The Giving Voice to Mothers Study: Inequity and Mistreatment During Pregnancy and Childbirth in the United States," *Reproductive Health* 16, no. 1 (June 2019), https://doi.org/10.1186/s12978-019-0729-2.

Listening to Mothers in California, California Health Care Foundation, 2018, https://www.chcf.org/publication/data-snapshot-listening-mothers-california/#related-links-and-downloads.

40. William A. Grobman et al., "Development of a Nomogram for Prediction of Vaginal Birth After Cesarean Delivery," *Obstetrics & Gynecology* 109, no. 4 (April 2007), https://doi.org/10.1097/01.aog.0000259312.36053.02.

Sally C. Curtin et al., "Maternal Morbidity for Vaginal and Cesarean Deliveries, According to Previous Cesarean History: New Data From the Birth Certificate, 2013," *National Vital Statistics Reports* 64, no. 4 (May 2015), 1–13, back cover, https://www.cdc.gov/nchs/data/nvsr/nvsr64/nvsr64_04.pdf.

41. Mark B. Landon et al., "The MFMU Cesarean Registry: Factors Affecting the Success of Trial of Labor After Previous Cesarean Delivery," *American Journal of Obstetrics & Gynecology* 193, no. 3 (September 2005), https://doi.org/10.1016/j.ajog.2005.05.066.

Annie L. Hollard et al., "Ethnic Disparity in the Success of Vaginal Birth after Caesarean Delivery," *Journal of Maternal-Fetal & Neonatal Medicine* 19, no. 8 (August 2006), https://doi.org/10.1080/14767050600847809.

Darshali A. Vyas et al., "Challenging the Use of Race in the Vaginal Birth after Cesarean Section Calculator," *Womens Health Issues* 29, no. 3 (May/June 2019), https://doi.org/10.1016/j.whi.2019.04.007.

42. William A. Grobman et al., "Prediction of Vaginal Birth after Cesarean Delivery in Term Gestations: A Calculator Without Race and Ethnicity," *American Journal of Obstetrics & Gynecology* 225, no. 6 (December 2021), https://doi.org/10.1016/j.ajog.2021.05.021.

Joyce A. Martin et al., "Births: Final Data for 2019," *National Vital Statistics Reports* 70, no. 2 (April 2021), http://dx.doi.org/10.15620/cdc:100472.

Sierra Washington et al., "Racial and Ethnic Differences in Indication for Primary Caesarean Delivery at Term: Experience at one US Institution," *Birth* 39, no. 2 (June 2012), https://doi.org/10.1111%2Fj.1523-536X.2012.00530.x.

43. Hannah Summers, "Guidance to Induce Minority Ethnic Pregnancies Earlier Condemned as Racist," *Guardian*, July 13, 2021, https://www.theguardian.com/global-development/2021/jul/13/nice-guidance-to-induce-minority-ethnic-pregnancies-earlier-condemned-as-racist.

Christine Douglass and Amali Lokugamage, "Racial Profiling for Induction of Labour: Improving Safety or Perpetuating Racism?," *BMJ*, November 2, 2021, https://doi.org/10.1136/bmj.n2562.

"Inducing Labour," National Institute for Health and Care Excellence, 2021, https://www.nice.org.uk/guidance/indevelopment/gid-ng10082/documents.

44. Katherine Kortsmit et al., "Abortion Surveillance—United States, 2019," *Morbidity and Mortality Weekly Report Surveillance Summary* 70, no. SS-9 (Summer 2021), http://dx.doi.org/10.15585/mmwr.ss7009a1.
    Samantha Artiga et al., "What Are the Implications of the Overturning of Roe v. Wade for Racial Disparities?," Kaiser Family Foundation, 2022, https://www.kff.org/racial-equity-and-health-policy/issue-brief/what-are-the-implications-of-the-overturning-of-roe-v-wade-for-racial-disparities/.
45. Siobhan Quenby et al., "Miscarriage Matters: The Epidemiological, Physical, Psychological, and Economic Costs of Early Pregnancy Loss," *Lancet* 397, no. 10285 (May 2021), https://doi.org/10.1016/S0140-6736(21)00682-6.
    Kacey Y. Eichelberger et al., "Black Lives Matter: Claiming a Space for Evidence-Based Outrage in Obstetrics and Gynecology," *American Journal of Public Health* 106, no. 10 (October 2016), https://doi.org/10.2105%2FAJPH.2016.303313.
    Ethnic Diversity in Fertility Treatment 2018, Human Fertilisation and Embryology Authority, 2021, https://www.hfea.gov.uk/about-us/publications/research-and-data/ethnic-diversity-in-fertility-treatment-2018/.
46. Christine Ekechi, "Addressing Inequality in Fertility Treatment," *Lancet* 398, no. 10301 (August 2021), https://doi.org/10.1016/s0140-6736(21)01743-8.
47. Harriet Grahame, "State Coroner's Court of New South Wales: Inquest into the Death of Naomi Williams," 2019.
48. *Maternal Deaths in Australia*, Australian Institute of Health and Welfare, 2020, http://doi.org/10.25816/7q4e-g697.
    Akilew A. Adane et al., "Socioethnic Disparities in Severe Maternal Morbidity in Western Australia: A Statewide Retrospective Cohort Study," *BMJ Open* 10, no. 11 (November 2020), https://doi.org/10.1136%2Fbmjopen-2020-039260.
49. Melinda Hayter and Rosie King, "Naomi Williams Inquest Concludes, with Coroner Calling for Change at NSW Hospital," Australian Broadcasting Corporation News, July 29, 2019, https://www.abc.net.au/news/2019-07-29/naomi-williams-tumut-sepsis-death-inquest-findings/11355244.
50. *Culturally Safe Health Care for Indigenous Australians*, Australian Institute of Health and Welfare, 2020, https://www.aihw.gov.au/reports/australias-health/culturally-safe-healthcare-indigenous-australians.
    Julianne Bryce, Elizabeth Foley, and Julie Reeves, "The Importance of Cultural Safety Not a Privilege," *Australian Nursing and Midwifery Journal* 25, no. 10 (May 2018).
51. Sue Kildea et al., "Effect of a Birthing on Country Service Redesign on Maternal and Neonatal Health Outcomes for First Nations Australians: A Prospective, Non-Randomised, Interventional Trial," *Lancet Global Health* 9, no. 5 (May 2021), https://doi.org/10.1016/s2214-109x(21)00061-9.
52. Natalie Dragon, "Birthing on Country: Improving Aboriginal & Torres Strait Islander infant and maternal health," *Australian Nursing and Midwifery Journal*, February 10, 2019, https://anmj.org.au/birthing-on-country-improving-indigenous-health/.
53. Vicki Van Wagner et al., "Reclaiming Birth, Health, and Community: Midwifery in the Inuit Villages of Nunavik, Canada," *Journal of Midwifery and Women's Health* 52, no. 4 (July/August 2007).
54. Margaret I. Rolfe et al., "The Distribution of Maternity Services Across Rural and Remote Australia: Does it Reflect Population Need?," *BMC Health Services Research* 17, no. 163 (2017), https://doi.org/10.1186/s12913-017-2084-8.
    Caroline S. E. Homer et al., "Maternal and Perinatal Outcomes by Planned Place of Birth in Australia 2000–2012: A Linked Population Data Study," *BMJ Open* 9, no. 10 (October 2019), https://doi.org/10.1 136/bmjopen-2019-029192.

Vanessa L. Scarf et al., "Maternal and Perinatal Outcomes by Planned Place of Birth Among Women with Low-Risk Pregnancies in High-Income Countries: A Systematic Review and Meta-Analysis," *Midwifery* 62 (July 2018), https://doi.org/10.1016/j.midw.2018.03.024.

55. Rachel R. Hardeman et al., "Roots Community Birth Center: A culturally- centered care model for improving value and equity in childbirth," *Healthcare* 8, no. 1 (March 2020), https://doi.org/10.1016/j.hjdsi.2019.100367.
56. Brigette Courtot et al., "Midwifery and Birth Centers Under State Medicaid Programs: Current Limits to Beneficiary Access to a High-Value Model of Care," *Milbank Quarterly* 98, no. 4 (December 2020), https://doi.org/10.1111/1468-0009.12473.
57. "Infant Mortality," Centers for Disease Control and Prevention, 2019, https://www.cdc.gov/reproductivehealth/maternalinfanthealth/infantmortality.htm.
   "Pregnancy and Ethnic Factors Influencing Births and Infant Mortality: 2013," Office for National Statistics, 2015, https://www.ons.gov.uk/peoplepopulationandcommunity/healthandsocialcare/causesofdeath/bulletins/pregnancyandethnicfactorsinfluencingbirthsandinfantmortality/2015-10-14.
   Yuan Huang et al., "Ethnicity and Maternal and Child Health Outcomes and Service Coverage in Western China: A Systematic Review and Meta-Analysis," *Lancet Global Health* 6, no. 1 (January 2018), https://doi.org/10.1016/S2214-109X(17)30445-X.
   Jayanta Kumar Bora, Rajesh Raushan, and Wolfgang Lutz, "The Persistent Influence of Caste on Under-Five Mortality: Factors that Explain the Caste-Based Gap in High Focus Indian States," *PLoS One* 14, no. 8 (2019), https://doi.org/10.1371%2Fjournal.pone.0211086.
58. Lynne Mofenson et al., *Recommendations of the US Public Health Service Task Force on the Use of Zidovudine to Reduce Perinatal Transmission of Human Immunodeficiency Virus*, 1994, https://www.cdc.gov/mmwr/preview/mmwrhtml/00032271.htm.
   Nathan Ford, Alexandra Calmy, and Edward J. Mills, "The First Decade of Antiretroviral Therapy in Africa," *Global Health* 7, no. 33 (September 2011), https://doi.org/10.1186/1744-8603-7-33.
   Bernhard Schwartländer, Ian Grubb, and Jos Perriëns, "The 10-Year Struggle to Provide Antiretroviral Treatment to People with HIV in the Developing World," *Lancet* 368, no. 9574 (August 2006), https://doi.org/10.1016/s0140-6736(06)69164-2.
   Philip Schellekens, *The Unfinished Business of COVID-19 Vaccination*, Pandem-ic, 2023, https://pandem-ic.com/the-unfinished-business-of-covid-19-vaccination/.
59. Didier Fassin and Helen Schneider, "The Politics of AIDS in South Africa: Beyond the Controversies," *BMJ* 326, no. 7387 (March 2003), https://doi.org/10.1136%2Fbmj.326. 7387.495.
   Chandré Gould and Peter I. Folb, "The South African Chemical and Biological Warfare Program: An Overview," *Nonproliferation Review*, Fall/Winter 2000.
60. Stephen J. Challacombe, "Global Inequalities in HIV Infection," *Oral Diseases* 26, suppl. 1 (September 2020), https://doi.org/10.1111/odi.13386.
   M. Mabaso et al., "HIV Prevalence in South Africa Through Gender and Racial Lenses: Results from the 2012 Population-Based National Household Survey," *International Journal for Equity in Health* 18, no. 167 (2019), https://doi.org/10.1186/s12939-019-1055-6.
61. "Racial inequities in HIV," *Lancet HIV* 7 (July 2020), https://doi.org/10.1016/S2352-3018(20)30173-9.
   "Impact on Racial and Ethnic Minorities," HIV.gov, 2022, https://www.hiv.gov/hiv-basics/overview/data-and-trends/impact-on-racial-and-ethnic-minorities.
   Ya-lin A. Huang et al., "HIV Preexposure Prophylaxis, by Race and Ethnicity—United States, 2014–2016," *Morbidity and Mortality Weekly Report* 67, no. 41 (October 19, 2018), http://dx.doi.org/10.15585/mmwr.mm6741a3.
62. Jeffrey S. Becasen et al., "Estimating the Prevalence of HIV and Sexual Behaviors among the US Transgender Population: A Systematic Review and Meta-Analysis, 2006–2017," *American Journal of Public Health* 109, no. 1 (January 2019), https://doi.org/10.2105/ajph.2018.304727.

Rageshri Dhairyawan et al., "Study for the UK Collaborative HIV Cohort: in HIV Clinical Outcomes Amongst Heterosexuals in the United Kingdom by Ethnicity," *AIDS* 35, no. 11 (September 2021), https://doi.org/10.1097/qad.0000000000002942.

63. "2020 Global AIDS Update—Seizing the Moment—Tackling Entrenched Inequalities to End Epidemics," UNAIDS, 2020, https://www.unaids.org/en/resources/documents/2020/global-aids-report.
    Becky L. Genberg et al., "A Comparison of HIV/AIDS-Related Stigma in Four Countries: Negative Attitudes and Perceived Acts of Discrimination Towards People Living with HIV/AIDS," *Social Science & Medicine* 68, no. 12 (June 2009), https://doi.org/10.1016/j.socscimed.2009.04.005.
64. "Uncovered: Online Hate Speech in the Covid Era," Ditch the Label, 2021, https://www.ditchthelabel.org/research-papers/hate-speech-report-2021/.
    Anne Saw, Aggie J. Yellow Horse, and Russell Jeung, *Stop AAPI Hate Mental Health Report*, 2021, https://stopaapihate.org/wp-content/uploads/2021/05/Stop-AAPI-Hate-Mental-Health-Report-210527.pdf.
    COVID-19 Racism Incident Report Survey Comprehensive Report 2021, Asian Australian Alliance, 2021, https://asianaustralianalliance.net/wp-content/uploads/2021/07/Comprehensive-AAA-Report.pdf.
    Narissa Subramoney, "Spanish Newspaper Apologises for Racist Omicron Cartoon," *The Citizen*, December 5, 2021, https://www.citizen.co.za/news/spanish-newspaper-apologises-racist-omicron-cartoon/.
    Michael Garthe, "DIE RHEINPFALZ im Shitstorm," *Die Rheinpfalz*, November 29, 2021, https://www.rheinpfalz.de/politik_artikel,-die-rheinpfalz-im-shitstorm-_arid,5285134.html.
65. Carolyn Smith-Morris, "Epidemiological Placism in Public Health Emergencies: Ebola in Two Dallas Neighborhoods," *Social Science & Medicine* 179 (April 2017), https://doi.org/10.1016/j.socscimed.2017.02.036.
    Fahimeh Saeed et al., "A Narrative Review of Stigma Related to Infectious Disease Outbreaks: What Can Be Learned in the Face of the Covid-19 Pandemic?," *Front Psychiatry* 11 (2020), https://doi.org/10.3389%2Ffpsyt.2020.565919.
66. Layal Liverpool, "An Unequal Society Means Covid-19 Is Hitting Ethnic Minorities Harder," *New Scientist*, April 21, 2020, https://www.newscientist.com/article/2241278-an-unequal-society-means-covid-19-is-hitting-ethnic-minorities-harder/.
    Aryati Yashadhana et al., "Indigenous Australians at Increased Risk of Covid-19 due to Existing Health and Socioeconomic Inequities," *Lancet Regional Health West Pacific* 1, no. 100007 (August 2020), https://doi.org/10.1016%2Fj.lanwpc.2020.100007.
    Pedro Baqui et al., "Ethnic and Regional Variations in Hospital Mortality from COVID-19 in Brazil: A Cross-Sectional Observational Study," *Lancet Global Health* 8, no. 8 (August 2020), https://doi.org/10.1016/S2214-109X(20)30285-0.
67. H. Zhao et al., "Ethnicity, Deprivation and Mortality due to 2009 Pandemic Influenza A(H1N1) in England during the 2009/2010 Pandemic and the First Post-Pandemic Season," *Epidemiology & Infection* 143, no. 16 (December 2015), https://doi.org/10.1017/s0950268815000576.
68. Grace A. Noppert et al., "Race and Nativity Are Major Determinants of Tuberculosis in the US: Evidence of Health Disparities in Tuberculosis Incidence in Michigan, 2004–2012," *BMC Public Health* 17, no. 1 (June 2017), https://doi.org/10.1186/s12889-017-4461-y.
    Sally Hayward et al., "Factors Influencing the Higher Incidence of Tuberculosis among Migrants and Ethnic Minorities in the UK," F1000Research 7 (2018), https://doi.org/10.12688%2Ff1000research.14476.2.
69. "BHF Live & Ticking April 2021—Inequalities and the Heart (live-streamed via YouTube)," British Heart Foundation, 2021, https://www.youtube.com/watch?v=HVXjCMyGQGU.

70. Nishi Chaturvedi, "Ethnic Differences in Cardiovascular Disease," *Heart* 89, no. 6 (June 2003), https://doi.org/10.1136/heart.89.6.681.
71. Rohini Mathur et al., "Ethnic Disparities in Initiation and Intensification of Diabetes Treatment in Adults with Type 2 Diabetes in the UK, 1990–2017: A Cohort Study," *PLoS Medicine* 17, no. 5 (May 2020), https://doi.org/10.1371/journal.pmed.1003106.
    Nikita Berman, Melvyn Mark Jones, and Daan A. De Coster, "'Just Like a Normal Pain, What do People with Diabetes Mellitus Experience when Having a Myocardial Infarction: A Qualitative Study Recruited from UK hospitals," *BMJ Open* 7, no. 9 (September 2017), https://doi.org/10.1136/bmjopen-2016-015736.
72. Aliki-Eleni Farmaki et al., "Type 2 Diabetes Risks and Determinants in Second-Generation Migrants and Mixed Ethnicity People of South Asian and African Caribbean Descent in the UK," *Diabetologia* 65, no. 1 (January 2022), https://doi.org/10.1007/s00125-021-05580-7.
73. Kosuke Inoue et al., "Urinary Stress Hormones, Hypertension, and Cardiovascular Events: The Multi-Ethnic Study of Atherosclerosis," *Hypertension* 78, no. 5 (November 2021), https://doi.org/10.1161/hypertensionaha.121.17618.
    Mei-Yan Liu et al., "Association Between Psychosocial Stress and Hypertension: A Systematic Review and Meta-Analysis," *Neurological Research* 39, no. 6 (June 2017), https://doi.org/10.1080/01616412.2017.1317904.
74. Mario Sims et al., "Perceived Discrimination and Hypertension Among African Americans in the Jackson Heart Study," *American Journal of Public Health* 102, suppl. 1 (May 2012), https://doi.org/10.2105/ajph.2011.300523.
    Margaret T. Hicken et al., "Racial/Ethnic Disparities in Hypertension Prevalence: Reconsidering the Role of Chronic Stress," *American Journal of Public Health* 104, no. 1 (January 2014), https://doi.org/10.2105%2FAJPH.2013.301395.
    Sula Mazimba and Pamela N. Peterson, "JAHA Spotlight on Racial and Ethnic Disparities in Cardiovascular Disease," *Journal of the American Heart Association* 10, no. 17 (September 2021), https://doi.org/10.1161/JAHA.121.023650.
75. Laura Benschop et al., "Future Risk of Cardiovascular Disease Risk Factors and Events in Women after a Hypertensive Disorder of Pregnancy," *Heart* 105, no. 16 (August 2019), https://doi.org/10.1136/heartjnl-2018-313453.
76. Angela N. Giaquinto et al., "Cancer Statistics for African American/Black People 2022," *CA: A Cancer Journal for Clinicians* 72, no. 3 (May 2022), https://doi.org/10.3322/caac.21718.
    "An Update on Cancer Deaths in the United States," Centers for Disease Control and Prevention, 2022, https://www.cdc.gov/cancer/dcpc/research/update-on-cancer-deaths/index.htm.
    "United States Cancer Statistics: Data Visualizations," Centers for Disease Control and Prevention, 2018, https://gis.cdc.gov/Cancer/USCS/#/Demographics/.
77. Michelle Tong, Latoya Hill, and Samantha Artiga, *Racial Disparities in Cancer Outcomes, Screening, and Treatment*, KFF, February 3, 2022, https://www.kff.org/racial-equity-and-health-policy/issue-brief/racial-disparities-in-cancer-outcomes-screening-and-treatment/.
    Aileen Bui et al., "Race, Ethnicity, and Socioeconomic Status Are Associated With Prolonged Time to Treatment After a Diagnosis of Colorectal Cancer: A Large Population-Based Study," *Gastroenterology* 160 (2021), https://doi.org/10.1053/j.gastro.2020.10.010.
    Prethibha George et al., "Diagnosis and Surgical Delays in African American and White Women with Early-Stage Breast Cancer," *Journal of Women's Health* 24, no. 3 (March 2015), https://doi.org/10.1089%2Fjwh.2014.4773.
78. Lisa J. Whop et al., "Towards Global Elimination of Cervical Cancer in All Groups of Women," *Lancet Oncology* 20, no. 5 (May 2019), https://doi.org/10.1016/s1470-2045(19)30237-2.
    Tedros Adhanom Ghebreyesus, "Improving the Health of Indigenous People Globally," *Lancet Oncology* 19, no. 6 (June 2018), https://doi.org/10.1016/S1470-2045(18)30375-9.
    Suzanne P. Moore et al., "Cancer Incidence in Indigenous People in Australia, New Zealand, Canada, and the USA: A Comparative Population-Based Study," *Lancet Oncology* 16, no. 15 (November 2015), https://doi.org/10.1016/s1470-2045(15)00232-6.

L. Jackson Pulver et al., "Indigenous Health: Australia, Canada, Aotearoa, New Zealand and the United States: Laying Claim to a Future that Embraces Health for Us All," World Health Report, 2010, https://researchdirect.westernsydney.edu.au/islandora/object/uws:35500.

79. "Eliminating Cervical Cancer," Victoria University of Wellington, https://www.wgtn.ac.nz/health/centres/national-centre-for-womens-health-research-aotearoa/eliminating-cervical-cancer.

80. Anna Adcock et al., "Acceptability of Self-Taken Vaginal HPV Sample for Cervical Screening among an Under-Screened Indigenous Population," *Australian and New Zealand Journal of Obstetrics and Gynaecology* 59, no. 2 (April 2019), https://doi.org/10.1111/ajo.12933.
    Evelyn Jane MacDonald et al., "Reaching Under-Screened/Never-Screened Indigenous Peoples with Human Papilloma Virus Self-Testing: A Community-Based Cluster Randomised Controlled Trial," *Australia and New Zealand Journal of Obstetrics and Gynaecology* 61, no. 1 (February 2021), https://doi.org/10.1111/ajo.13285.

81. Tanimola Martins et al., "Assessing Ethnic Inequalities in Diagnostic Interval of Common Cancers: A Population-Based UK Cohort Study," *Cancers (Basel)* 14, no. 13 (June 2022), https://doi.org/10.3390/cancers14133085.

82. Kimberly A. Terrell and Gianna St. Julien, "Air Pollution Is Linked to Higher Cancer Rates among Black or Impoverished Communities in Louisiana," *Environmental Research Letters* 17, no. 1 (2022), http://doi.org/10.1088/1748-9326/ac4360.
    Daniela Fecht et al., "Associations Between Air Pollution and Socioeconomic Characteristics, Ethnicity and Age Profile of Neighbourhoods in England and the Netherlands," *Environmental Pollution* 198 (2015), https://doi.org/10.1016/j.envpol.2014.12.014.
    Hong-Bae Kim et al., "Long-Term Exposure to Air Pollutants and Cancer Mortality: A Meta-Analysis of Cohort Studies," *International Journal of Environmental Research and Public Health* 15, no. 11 (November 2018), https://doi.org/10.3390/ijerph15112608.
    P. Mudu et al., "The Importance and Challenge of Carcinogenic Air Pollutants for Health Risk and Impact Assessment," *European Journal of Public Health* 30, suppl. 5 (2020), https://doi.org/10.1093/eurpub/ckaa165.841.
    Laurel A. Schaider et al., "Environmental Justice and Drinking Water Quality: Are there Socioeconomic Disparities in Nitrate Levels in US Drinking Water?," *Environmental Health* 18, no. 3 (2019), https://doi.org/10.1186/s12940-018-0442-6.
    Hee-Soon Juon et al., "Racial Disparities in Occupational Risks and Lung Cancer Incidence: Analysis of the National Lung Screening Trial," *Journal of Preventive Medicine* 143 (2021), https://doi.org/10.1016/j.ypmed.2020.106355.

83. "Der Report 2020," Afrozensus, 2021, https://afrozensus.de/reports/2020/.

84. Anjum Memon et al., "Perceived Barriers to Accessing Mental Health Services among Black and Minority Ethnic (BME) Communities: A Qualitative Study in Southeast England," *BMJ Open* 6, no. 11 (November 2016), https://doi.org/10.1136/bmjopen-2016-012337.
    Oanh L. Meyer and Nolan Zane, "The Influence of Race and Ethnicity in Clients' Experiences of Mental Health Treatment," *Journal of Community Psychology* 41, no. 7 (2013), https://doi.org/10.1002%2Fjcop.21580.
    Jude Mary Cénat, "How to Provide Anti-racist Mental Health Care," *Lancet Psychiatry* 7, no. 11 (November 2020), https://doi.org/10.1016/s2215-0366(20)30309-6.
    Emmanuel M. Ngui et al., "Mental Disorders, Health Inequalities and Ethics: A Global Perspective," *International Review of Psychiatry* 22, no. 3 (2010), https://doi.org/10.3109%2F09540261.2010.485273.

85. Ozlem Eylem et al., "Stigma for Common Mental Disorders in Racial Minorities and Majorities a Systematic Review and Meta-Analysis," *BMC Public Health* 20 (2020), https://doi.org/10.1186/s12889-020-08964-3.

86. Stephanie Anne Shelton and Aryah O. S. Lester, "A Narrative Exploration of the Importance of Intersectionality in a Black Trans Woman's Mental Health Experiences," *International*

Journal of Transgender Health 23, no. 1–2 (2020), https://doi.org/10.1080/26895269.2020.1838393.

87. Sarah Ketchen Lipson et al., "Mental Health Disparities Among College Students of Color," *Journal of Adolescent Health* 63, no. 3 (September 2018), https://doi.org/10.1016/j.jadohealth.2018.04.014.
    Janet R. Cummings et al., "Racial and Ethnic Differences in Minimally Adequate Depression Care Among Medicaid-Enrolled Youth," *Journal of the American Academy of Child & Adolescent Psychiatry* 58, no. 1 (January 2019), https://doi.org/10.1016/j.jaac.2018.04.025.
    Alfiee M. Breland-Noble, "Mental Healthcare Disparities Affect Treatment of Black Adolescents," *Psychiatric Annals* 34, no. 7 (July 2004), https://doi.org/10.3928%2F0048-5713-20040701-14.
88. "Mental Health Data Dashboard," NYC Mayor's Office of Community Mental Health, https://mentalhealth.cityofnewyork.us/dashboard/.
89. Gargie Ahmad et al., "Prevalence of Common Mental Disorders and Treatment Receipt for People from Ethnic Minority Backgrounds in England: Repeated Cross-Sectional Surveys of the General Population in 2007 and 2014," *British Journal of Psychiatry* 221, no. 3 (September 2022), https://doi.org/10.1192/bjp.2021.179.
    Phoebe Barnett et al., "Ethnic variations in compulsory detention under the Mental Health Act: a systematic review and meta-analysis of international data," *Lancet Psychiatry* 6, no. 4 (April 2019), https://doi.org/10.1016/s2215-0366(19)30027-6.
    Sarah Rosenfield, "Race Differences in Involuntary Hospitalization: Psychiatric vs. Labeling Perspectives," *Journal of Health and Social Behavior* 25, no. 1 (March 1984), https://psycnet.apa.org/doi/10.2307/2136701.
    Michael A. Gara et al., "A Naturalistic Study of Racial Disparities in Diagnoses at an Outpatient Behavioral Health Clinic," *Psychiatric Services* 70, no. 2 (February 2018), https://doi.org/10.1176/appi.ps.201800223.
    Stephen M. Strakowski et al., "Ethnicity and Diagnosis in Patients with Affective Disorders," *Journal of Clinical Psychiatry* 64, no. 64 (July 2003), https://doi.org/10.4088/jcp.v64n0702.
    Rebecca Pinto, Mark Ashworth, and Roger Jones, "Schizophrenia in Black Caribbeans Living in the UK: An Exploration of Underlying Causes of the High Incidence Rate," *British Journal of General Practice* 58, no. 551 (June 2008), https://doi.org/10.3399%2Fbjgp08X299254.
90. Ayana Jordan, Aza Stephen Allsop, Pamela Y. Collins, "Decriminalising being Black with Mental Illness," *Lancet Psychiatry* 8, no. 1 (January 2021), https://doi.org/10.1016/s2215-0366(20)30519-8.
    Marilyn D. Thomas, Nicholas P. Jewell, and Amani M. Allen, "Black and Unarmed: Statistical Interaction Between Age, Perceived Mental Illness, and Geographic Region among Males Fatally Shot by Police Using Case-Only Design," *Annals of Epidemiology* 53, 42-49.e3 (2021), https://doi.org/10.1016/j.annepidem.2020.08.014.
    Derecka Purnell, *Becoming Abolitionists: Police, Protests, and the Pursuit of Freedom* (New York: Astra House, 2021).
    Jacob Bor et al., "Police Killings and their Spillover Effects on the Mental Health of Black Americans: A Population-Based, Quasi-Experimental Study," *Lancet* 392, no. 10144 (July 2018), https://doi.org/10.1016/s0140-6736(18)31130-9.
91. Samuel A. Cartwright, "Report on the Diseases and Physical Peculiarities of the Negro Race," *New Orleans Medical and Surgical Journal* (1851).
92. W. Bromberg and F. Simon, "The 'Protest' Psychosis: A Special Type of Reactive Psychosis," *Archives of General Psychiatry* 19, no. 2 (August 1968), https://doi.org/10.1001/archpsyc.1968.01740080027005.
    Jonathan Metzl, *The Protest Psychosis: How Schizophrenia Became a Black Disease* (Boston: Beacon Press, 2009).
93. Helena Mendes Constante, Gerson Luiz Marinho, and João Luiz Bastos, "The Door Is Open, but Not Everyone May Enter: Racial Inequities in Healthcare Access Across Three

Brazilian Surveys," *Ciência & Saúde Coletiva* 26, no. 9 (September 2021), https://doi.org/10.1590/1413-81232021269.47412020.

Thomas Hone et al., "Association Between Expansion of Primary Healthcare and Racial Inequalities in Mortality Amenable to Primary Care in Brazil: A National Longitudinal Analysis," *PLoS Medicine*, 14, no. 5 (2017), https://doi.org/10.1371/journal.pmed.1002306.

94. Dom Phillips, "'Enormous Disparities': Coronavirus Death Rates Expose Brazil's Deep Racial Inequalities," *Guardian*, June 9, 2020, https://www.theguardian.com/world/2020/jun/09/enormous-disparities-coronavirus-death-rates-expose-brazils-deep-racial-inequalities.

95. Bianca Muniz, Bruno Fonseca, and Rute Pina, "Em duas semanas, número de negros mortos por coronavírus é cinco vezes maior no Brasil," *Pública*, May 6, 2020, https://apublica.org/2020/05/em-duas-semanas-numero-de-negros-mortos-por-coronavirus-e-cinco-vezes-maior-no-brasil/.

96. Edna Maria de Araújo et al., "Morbimortalidade pela Covid-19 segundo raça/cor/etnia: a experiência do Brasil e dos Estados Unidos," *Saúde em Debate* 44 (2020), https://doi.org/10.1590/0103-11042020E412.

97. Kia Lilly Caldwell and Edna Maria de Araújo, "COVID-19 Is Deadlier for Black Brazilians, A Legacy of Structural Racism that Dates Back to Slavery," *Conversation*, June 11, 2020, https://theconversation.com/covid-19-is-deadlier-for-black-brazilians-a-legacy-of-structural-racism-that-dates-back-to-slavery-139430.

Edward Telles, "Racial Discrimination and Miscegenation," *UN Chronicle* 44, no. 3 (2008), https://www.un.org/en/chronicle/article/racial-discrimination-and-miscegenation-experience-brazil.

Valerie Wilson and William Darity Jr., "Understanding Black-White Disparities in Labor Market Outcomes Requires Models that Account for Persistent Discrimination and Unequal Bargaining Power," Economic Policy Institute, March 25, 2022, https://www.epi.org/unequalpower/publications/understanding-black-white-disparities-in-labor-market-outcomes/.

98. Philippa Stroud, "Measuring Poverty Before the Covid-19 Pandemic," Social Metrics Commission, 2021, https://socialmetricscommission.org.uk/measuring-poverty-before-the-covid-19-pandemic/.

Edward P. Havranek, "The Influence of Social and Economic Factors on Heart Disease," *JAMA Cardiology* 4, no. 12 (2019), https://doi.org/10.1001/jamacardio.2019.3802.

Silvia Stringhini et al., "Association of Lifecourse Socioeconomic Status with Chronic Inflammation and Type 2 Diabetes Risk: The Whitehall II Prospective Cohort Study," *PLoS Medicine* 10, no. 7 (2013), https://doi.org/10.1371/journal.pmed.1001479.

99. Craig Evan Pollack et al., "Should Health Studies Measure Wealth? A Systematic Review," *American Journal of Preventive Medicine* 33, no. 3 (September 2007), https://doi.org/10.1016/j.amepre.2007.04.033.

Craig Evan Pollack et al., "Do wealth disparities contribute to health disparities within racial/ethnic groups?," *Journal of Epidemiology and Community Health* no. 5 (May 2013), https://doi.org/10.1136/jech-2012-200999.

Elizabeth Sweet et al., "The High Price of Debt: Household Financial Debt and Its Impact on Mental and Physical Health," *Social Science & Medicine* 91 (2013), https://doi.org/10.1016/j.socscimed.2013.05.009.

100. Terry Gross, "'Forgotten History' of How the US Government Segregated America," NPR, May 3, 2017, https://www.npr.org/2017/05/03/526655831/a-forgotten-history-of-how-the-u-s-government-segregated-america.

Cheryl W. Thompson et al., "Racial Covenants, a Relic of the Past, Are Still on the Books Across the Country," NPR, November 17, 2021, https://www.npr.org/2021/11/17/1049052531/racial-covenants-housing-discrimination.

101. Darrick Hamilton, "Post-Racial Rhetoric, Racial Health Disparities, and Health Disparity Consequences of Stigma, Stress, and Racism," Washington Center for Equitable Growth, October 10, 2017, https://equitablegrowth.org/working-papers/racial-health-disparities/.
102. "Who's Working in HE?: Personal Characteristics," Higher Education Statistics Agency, February 21, 2023, https://www.hesa.ac.uk/data-and-analysis/staff/working-in-he/characteristics.
103. Maria Fatima Marinho et al., "Racial Disparity in Excess Mortality in Brazil During COVID-19 Times," *European Journal of Public Health* 32, no. 1 (February 2022), https://doi.org/10.1093/eurpub/ckab097.
"The Unequal Impact of COVID-19: A Spotlight on Frontline Workers, Migrants and Racial/Ethnic Minorities," Organisation for Economic Co-operation and Development, March 17, 2022, https://www.oecd.org/coronavirus/policy-responses/the-unequal-impact-of-covid-19-a-spotlight-on-frontline-workers-migrants-and-racial-ethnic-minorities-f36e931e/.
Layal Liverpool, "Why Coronavirus Hit People From BAME Communities So Hard," *Wired*, August 25, 2020, https://www.wired.co.uk/article/bame-communities-coronavirus-uk.
Molly Kinder and Laura Stateler, "Essential Workers Comprise about Half of All Workers in Low Paid Occupations. They Deserve a $15 Minimum Wage," Brookings Institution, February 25, 2021, https://www.brookings.edu/blog/the-avenue/2021/02/05/essential-workers-deserve-minimum-wage-increase/.
104. Claire Marshall, "Rosamund Adoo-Kissi-Debrah: 'Did Air Pollution Kill My Daughter?,'" BBC, November 29, 2020, https://www.bbc.com/news/stories-55106501.
105. Philip Barlow, "Annex A: Regulation 28: Report to Prevent Future Deaths (1)," coroner's investigation into death of Ella Kissi-Debrah, April 20, 2021, https://www.judiciary.uk/wp-content/uploads/2021/04/Ella-Kissi-Debrah-2021-0113-1.pdf.
"The Health Effects of Air Pollution: Driving Policy and Legislative Change to Improve Air Quality and Protect Public Health," University of Southampton, https://www.southampton.ac.uk/medicine/research/impact/health-effects-of-air-pollution.page.
Adam Vaughan, "Landmark Ruling Says Air Pollution Contributed to Death of 9-Year-Old," *New Scientist*, December 16, 2020, https://www.newscientist.com/article/2263165-landmark-ruling-says-air-pollution-contributed-to-death-of-9-year-old/.
106. Adam Vaughan, "Rosamund Kissi-Debrah: Clean Air 'Ella's Law' Would Honour Her Memory," *New Scientist*, December 18, 2020, https://www.newscientist.com/article/2263428-rosamund-kissi-debrah-clean-air-ellas-law-would-honour-her-memory/.
107. Gopalakrishnan Netuveli et al., "Ethnic Variations in UK Asthma Frequency, Morbidity, and Health-Service Use: A Systematic Review and Meta-Analysis," *Lancet* 365, no. 9456 (January 2005), https://doi.org/10.1016/s0140-6736(05)17785-x.
Adam Vaughan, "London's Black Communities Disproportionately Exposed to Air Pollution—Study," *Guardian*, October 11, 2016, https://www.theguardian.com/environment/2016/oct/10/londons-black-communities-disproportionately-exposed-to-air-pollution-study.
108. "Which Neighbourhoods Have the Worst Air Pollution?," Friends of the Earth, October 21, 2022, https://policy.friendsoftheearth.uk/insight/which-neighbourhoods-have-worst-air-pollution.
109. Christopher W. Tessum et al., "Inequity in Consumption of Goods and Services Adds to Racial–Ethnic Disparities in Air Pollution Exposure," *Proceedings of the National Academy of Sciences* 116, no. 13 (March 2019), https://doi.org/10.1073/pnas.1818859116.
"Asthma and African Americans," US Department of Health and Human Services Office of Minority Health, 2021, https://minorityhealth.hhs.gov/omh/browse.aspx?lvl=4&lvlid=15.
110. Ker Than, "Can California Farmers Save Water and the Dying Salton Sea?," *National Geographic*, February 18, 2014, https://www.nationalgeographic.com/history/article/140218-salton-sea-imperial-valley-qsa-water-conservation.
Ryan Llamas, "What's Wrong with the Salton Sea?," National Audubon Society, April 29, 2020, https://ca.audubon.org/news/whats-wrong-salton-sea.

111. Yoshira Ornelas Van Horne et al., "Influences of Windblown Particulate Matter on Children's Respiratory Health Residing Near the Salton Sea, California," ISEE Conference Abstracts, 2021, no. 1 (2021), https://ehp.niehs.nih.gov/doi/abs/10.1289/isee.2021.P-491.
Shohreh F. Farzan et al., "Assessment of Respiratory Health Symptoms and Asthma in Children near a Drying Saline Lake," *International Journal of Environmental Research and Public Health* 16, no. 20 (October 2019), https://doi.org/10.3390/ijerph16203828.
112. Jill E. Johnston et al., "The Disappearing Salton Sea: A Critical Reflection on the Emerging Environmental Threat of Disappearing Saline Lakes and Potential Impacts on Children's Health," *Science of the Total Environment* 663 (May 2019), https://doi.org/10.1016/j.scitotenv.2019.01.365.
113. Sudipta Sarkar et al., "Impact of Deadly Dust Storms (May 2018) on Air Quality, Meteorological, and Atmospheric Parameters Over the Northern Parts of India," *GeoHealth* 3, no. 3 (March 2019), https://doi.org/10.1029/2018GH000170.
Jayajit Chakraborty and Pratyusha Basu, "Air Quality and Environmental Injustice in India: Connecting Particulate Pollution to Social Disadvantages," *International Journal of Environmental Research and Public Health* 18, no. 1 (January 2021), preprint at https://doi.org/10.3390/ijerph18010304.
114. *Toxic Wastes and Race in the United States: A National Report on the Racial and Socio-Economic Characteristics of Communities With Hazardous Waste Sites*, United Church of Christ Commission for Racial Justice, 1987, https://www.ucc.org/wp-content/uploads/2020/12/ToxicWastesRace.pdf.
Robert D. Bullard, *Dumping in Dixie: Race, Class, and Environmental Quality* (Routledge, 1990).
Robert D. Bullard et al., "Toxic Wastes and Race at Twenty: Why Race Still Matters After all of these Years," *Environmental Law* 38, no. 2 (Spring 2008), http://www.jstor.org/stable/43267204.
Paul Mohai and Robin Saha, "Which Came First, People or Pollution? Assessing the Disparate Siting and Post-Siting Demographic Change Hypotheses of Environmental Injustice," *Environmental Research Letters* 10, no. 11 (2015), http://doi.org/10.1088/1748-9326/10/11/115008.
Michael Mascarenhas, Ryken Grattet, and Kathleen Mege, "Toxic Waste and Race in Twenty-First Century America: Neighborhood Poverty and Racial Composition in the Siting of Hazardous Waste Facilities," *Environment and Society* 12, no. 1 (September 2021), http://dx.doi.org/10.3167/ares.2021.120107.
115. Kristi Pullen Fedinick, Steve Taylor, and Michele Roberts, *Watered Down Justice,* Natural Resources Defense Council, September 2019, https://www.nrdc.org/resources/watered-down-justice.
116. Mona Hanna-Attisha et al., "Elevated Blood Lead Levels in Children Associated with the Flint Drinking Water Crisis: A Spatial Analysis of Risk and Public Health Response," *American Journal of Public Health* 106, no. 2 (February 2016), https://doi.org/10.2105%2FAJPH.2015.303003.
Deniz Yeter, Ellen C. Banks, and Michael Aschner, "Disparity in Risk Factor Severity for Early Childhood Blood Lead among Predominantly African-American Black Children: The 1999 to 2010 US NHANES," *International Journal of Environmental Research and Public Health* 17, no. 5 (February 2020), https://doi.org/10.3390/ijerph17051552.
117. "USA: Environmental Racism in 'Cancer Alley' Must End—Experts," United Nations, March 2, 2021, https://www.ohchr.org/en/press-releases/2021/03/usa-environmental-racism-cancer-alley-must-end-experts.
118. Johnnye Lewis, Joseph Hoover, and Debra MacKenzie, "Mining and Environmental Health Disparities in Native American Communities," *Current Environmental Health Reports* 4, no. 2 (2017), https://doi.org/10.1007/s40572-017-0140-5.

Yoshira Ornelas Van Horne et al., "Impacts to Diné Activities with the San Juan River after the Gold King Mine Spill," *Journal of Exposure Science & Environmental Epidemiology* 31, no. 5 (September 2021), https://doi.org/10.1038/s41370-021-00290-z.
Yoshira Ornelas Van Horne, "Exposures and Health Risks of the Diné Communities Impacted by the Gold King Mine Spill" (PhD diss., University of Arizona, 2019), http://hdl.handle.net/10150/636668.

119. Aline Philibert, Myriam Fillion, and Donna Mergler, "Mercury Exposure and Premature Mortality in the Grassy Narrows First Nation Community: A Retrospective Longitudinal Study," *Lancet Planetary Health* 4, no. 4 (April 2020).
Jack Graham, "Canada Votes To Collect Data to Document 'Environmental Racism,'" Reuters, March 25, 2021, https://www.reuters.com/article/us-canada-environment-racism-trfn-idUSKBN2BG3BV.
"Amazonian Indigenous Peoples Demand Justice One Year After Ecuador Oil Spill," Amazon Watch, April 7, 2021, https://amazonwatch.org/news/2021/0407-amazonian-indigenous-peoples-demand-justice-one-year-after-ecuador-oil-spill.

120. S. Nazrul Islam and John Winkel, "Climate Change and Social Inequality," United Nations, 2017, https://digitallibrary.un.org/record/3859027?ln=en.
Sabine Minninger, Laura Schäfer, and Vera Künzel, "Building Resilience: Climate Impacts and Corona," Germanwatch, April 14, 2020, https://www.germanwatch.org/en/18535.

121. Renata R. Silva et al., "Basic Sanitation: A New Indicator for the Spread of COVID-19?," *Transactions of the Royal Society of Tropical Medicine and Hygiene*, 115, no. 7 (July 2021), https://doi.org/10.1093/trstmh/traa187.
Marina Soltan et al., "To What Extent Are Social Determinants of Health, Including Household Overcrowding, Air Pollution and Housing Quality Deprivation, Modulators of Presentation, ITU Admission and Outcomes Among Patients with SARS-COV-2 Infection in an Urban Catchment Area in Birmingham, United Kingdom?," *Thorax* 76, suppl. 1 (February 2021), http://dx.doi.org/10.1136/thorax-2020-BTSabstracts.414.
Eric B. Brandt, Andrew F. Beck, and Tesfaye B. Mersha, "Air Pollution, Racial Disparities, and COVID-19 Mortality," *Journal of Allergy and Clinical Immunology* 146, no. 1 (July 2020), https://doi.org/10.1016%2Fj.jaci.2020.04.035.
X. Wu et al., "Air Pollution and COVID-19 Mortality in the United States: Strengths and Limitations of an Ecological Regression Analysis," *Science Advances* 6, no. 45 (November 2020), https://doi.org/10.1126/sciadv.abd4049.
Tanujit Dey and Francesca Dominici, "COVID-19, Air Pollution, and Racial Inequity: Connecting the Dots," *Chemical Research in Toxicology* 34, no. 3 (March 2021), https://doi.org/10.1021/acs.chemrestox.0c00432.

122. Lina Toleikyte and Sarah Salway, *Local Action on Health Inequalities: Understanding and Reducing Ethnic Inequalities in Health* (London: Public Health England, 2018), https://www.ucl.ac.uk/epidemiology-health-care/news/2018/sep/local-action-health-inequalities-understanding-and-reducing-ethnic-inequalities-health.
Renee E. Walker, Christopher R. Keane, and Jessica G. Burke, "Disparities and Access To Healthy Food in the United States: A Review of Food Deserts Literature," *Health & Place* 16, no. 5 (September 2010), https://doi.org/10.1016/j.healthplace.2010.04.013.
Harriet A. Washington, *A Terrible Thing to Waste: Environmental Racism and Its Assault on the American Mind* (New York: Little, Brown and Company, 2019).
Harriet A. Washington, "How Environmental Racism Is Fuelling the Coronavirus Pandemic," *Nature* 581, no. 7808 (May 2020), https://doi.org/10.1038/d41586-020-01453-y.
Olivia Souza Honório et al., "Social Inequalities in the Surrounding Areas of Food Deserts and Food Swamps in a Brazilian Metropolis," *International Journal for Equity in Health* 20, no. 168 (2021), https://doi.org/10.1186/s12939-021-01501-7.
Kimberly Morland, Ana V. Diez Roux, and Steve Wing, "Supermarkets, Other Food Stores, and Obesity: The Atherosclerosis Risk in Communities Study," *American Journal*

of Preventive Medicine 30, no. 4 (April 2006), https://doi.org/10.1016/j.amepre.2005.11.003.

Susan H. Babey et al., "Designed for Disease: The Link Between Local Food Environments and Obesity and Diabetes," UCLA Center for Health Policy Research, April 2008, https://escholarship.org/uc/item/7sf9t5wx.

Craig M. Hales et al., "Prevalence of Obesity Among Adults and Youth: United States, 2015–2016," *NCHS Data Briefs* 288 (October 2017), https://www.cdc.gov/nchs/data/databriefs/db288.pdf.

"Overweight Adults," Ethnicity Facts and Figures, UK Office for Health Improvement and Disparities, March 24, 2023, https://www.ethnicity-facts-figures.service.gov.uk/health/diet-and-exercise/overweight-adults/latest.

123. Viniece Jennings and Cassandra Johnson Gaither, "Approaching Environmental Health Disparities and Green Spaces: An Ecosystem Services Perspective," *International Journal of Environmental Research and Public Health* 12, no. 2 (February 2015), https://doi.org/10.3390%2Fijerph120201952.

Research summary: Urban Green Nation; Building the Evidence Base, Design Council, 2010, https://www.designcouncil.org.uk/resources/report/urban-green-nation.

124. Rashawn Ray, "Black People Don't Exercise in My Neighborhood: Perceived Racial Composition and Leisure-Time Physical Activity Among Middle Class Blacks and Whites," *Social Science Research* 66 (August 2017), https://doi.org/10.1016/j.ssresearch.2017.03.008.

Helder Ferreira and Milena Karla Soares, "Violência e segurança pública: uma síntese da produção da Diest nos últimos dez anos," *Boletim de Analíse Político-Institucional* 29 (June 2021), https://repositorio.ipea.gov.br/bitstream/11058/10644/1/bapi_29_violencia_seguranca.pdf.

125. Clint Smith, "What a Racist Slur Does to the Body," *Atlantic*, October 22, 2022, https://www.theatlantic.com/ideas/archive/2022/10/racism-effects-health-black-americans/671822/.

126. "Why Covid-19 Is Killing Black People," *The United States of Anxiety*, WNYC, April 25, 2020, https://www.wnycstudios.org/podcasts/anxiety/episodes/why-covid-black-people.

127. Arline T. Geronimus, "The Weathering Hypothesis and the Health of African-American Women and Infants: Evidence and Speculations," *Ethnicity & Disease* 2, no. 3 (Summer 1992), https://www.jstor.org/stable/45403051.

128. "Covid-19 Has Shone a Light on Racial Disparities in Health," *The Economist*, November 21, 2020, https://www.economist.com/international/2020/11/21/covid-19-has-shone-a-light-on-racial-disparities-in-health.

129. Maya Dominguez McNeilly et al., "Effects of Racist Provocation and Social Support on Cardiovascular Reactivity in African American Women," *International Journal of Behavioral Medicine* 2, no. 4 (1995), https://doi.org/10.1207/s15327558ijbm0204_3.

C. Y. Fang and H. F. Myers, "The Effects of Racial Stressors and Hostility on Cardiovascular Reactivity in African American and Caucasian Men," *Health Psychology* 20, no. 1 (January 2001), https://doi.org/10.1037//0278-6133.20.1.64.

Max Guyll, Karen A. Matthews, and Joyce T. Bromberger, "Discrimination and Unfair Treatment: Relationship to Cardiovascular Reactivity Among African American and European American Women," *Health Psychology* 20, no. 5 (September 2001), https://doi.org/10.1037//0278-6133.20.5.315.

Bruce S. McEwen and Eliot Stellar, "Stress and the Individual. Mechanisms Leading to Disease," *Archives of Internal Medicine* 153, no. 18 (September 1993), https://doi.org/10.1001/archinte.1993.00410180039004.

Arline T. Geronimus et al., "'Weathering' and Age Patterns of Allostatic Load Scores Among Blacks and Whites in the United States," *American Journal of Public Health* 96, no. 5 (May 2006), https://doi.org/10.2105%2FAJPH.2004.060749.

M. Kristen Peek et al., "Allostatic Load Among Non-Hispanic Whites, Non-Hispanic Blacks, and People of Mexican Origin: Effects of Ethnicity, Nativity, and Acculturation," *American Journal of Public Health* 100, no. 5 (May 2010), https://doi.org/10.2105%2FAJPH.2007.129312.

Arline T. Geronimus et al., "Do US Black Women Experience Stress-Related Accelerated Biological Aging?: A Novel Theory and First Population-Based Test of Black-White Differences in Telomere Length," *Human Nature* 21, no. 1 (March 2010), https://doi.org/10.1007%2Fs12110-010-9078-0.

130. David R. Williams, "Miles to Go Before We Sleep: Racial Inequities in Health," *Journal of Health and Social Behavior* 53, no. 3 (September 2012), https://doi.org/10.1177%2F0022146512455804.

131. Kenneth D. Kochanek et al., "Deaths: Final Data for 2017," *National Vital Statistics Reports* 68, no. 9 (June 2019), https://www.cdc.gov/nchs/data/nvsr/nvsr68/nvsr68_09-508.pdf.

Theresa Andrasfay and Noreen Goldman, "Reductions in 2020 US Life Expectancy Due to COVID-19 and the Disproportionate Impact on the Black and Latino Populations," *Proceedings of the National Academy of Sciences* 118, no. 5 (January 2021), https://doi.org/10.1073/pnas.2014746118.

David R. Williams, "How Racism Makes us Sick," TEDMED talk, 2016, https://www.tedmed.com/talks/show?id=621421.

David R. Williams et al., "Racial Differences in Physical and Mental Health: Socio-economic Status, Stress and Discrimination," *Journal of Health Psychology* 2, no. 3 (July 1997), https://doi.org/10.1177/135910539700200305.

Allie Slemon et al., "Analysis of the Social Consequences and Value Implications of the Everyday Discrimination Scale (EDS): Implications for Measurement of Discrimination in Health Research," *Health Sociology Review* 31, no. 3 (November 2022), https://doi.org/10.1080/14461242.2021.1969980.

Ricci Harris et al., "Racism and Health: The Relationship between Experience of Racial Discrimination and Health in New Zealand," *Social Science & Medicine* 63, 1428–1441 (2006), https://doi.org/10.1016/j.socscimed.2006.04.009.

David H. Chae et al., "Do Experiences of Racial Discrimination Predict Cardiovascular Disease among African American Men? The Moderating Role of Internalized Negative Racial Group Attitudes," *Social Science & Medicine* 71, 1182–1188 (2010), https://doi.org/10.1016/j.socscimed.2010.05.045.

Teletia R. Taylor et al., "Racial Discrimination and Breast Cancer Incidence in US Black Women: The Black Women's Health Study," *American Journal of Epidemiology* 166, 46–54 (2007), https://doi.org/10.1093/aje/kwm056.

Shawn O. Utsey, "Racism and the Psychological Well-Being of African American Men," *Journal of African American Men*, 69–87 (1997), https://doi.org/10.1007/s12111-997-1011-1.

132. Pamela J. Sawyer et al., "Discrimination and the Stress Response: Psychological and Physiological Consequences of Anticipating Prejudice in Interethnic Interactions," *American Journal of Public Health* 102, no. 5 (May 2012), https://doi.org/10.2105%2FAJPH.2011.300620.

Layal Liverpool, "Systemic Racism: What Research Reveals about the Extent of its Impact," *New Scientist*, November 18, 2020, https://www.newscientist.com/article/mg24833093-900-systemic-racism-what-research-reveals-about-the-extent-of-its-impact/.

133. Thomas H. Holmes and Richard H. Rahe, "The Social Readjustment Rating Scale," *Journal of Psychosomatic Research* 11, no. 2 (1967), https://doi.org/10.1016/0022-3999(67)90010-4.

Shawn O. Utsey et al., "Cultural, Sociofamilial, and Psychological Resources that Inhibit Psychological Distress in African Americans Exposed to Stressful Life Events and Race-Related Stress," *Journal of Counseling Psychology* 55, no. 1 (2008), https://doi.org/10.1037/0022-0167.55.1.49.

Shawn O. Utsey and Joseph G. Ponterotto, "Development and Validation of the Index of Race-Related Stress (IRRS)," *Journal of Counseling Psychology* 43, no. 4 (1996), https://doi.org/10.1037/0022-0167.43.4.490.

Shawn O. Utsey et al., "Effect of Ethnic Group Membership on Ethnic Identity, Race-Related Stress, and Quality of Life," *Cultural Diversity and Ethnic Minority Psychology* 8, no. 4 (November 2002), https://doi.org/10.1037/1099-9809.8.4.367.

Shawn O. Utsey et al., "Race-Related Stress, Quality of Life Indicators, and Life Satisfaction among Elderly African Americans," *Cultural Diversity and Ethnic Minority Psychology* 8, no. 3 (August 2002), https://doi.org/10.1037/1099-9809.8.3.224.

Kathryn Freeman Anderson, "Diagnosing Discrimination: Stress from Perceived Racism and the Mental and Physical Health Effects," *Sociological Inquiry* 83, no. 1 (2013), https://doi.org/10.1111/j.1475-682X.2012.00433.x.

Cynthia M. Dolezsar et al., "Perceived Racial Discrimination and Hypertension: A Comprehensive Systematic Review," *Health Psychology* 33, no. 1 (January 2014), https://doi.org/10.1037/a0033718.

134. Danielle L. Beatty Moody et al., "Lifetime Discrimination Burden, Racial Discrimination, and Subclinical Cerebrovascular Disease among African Americans," *Health Psychology* 38, no. 1 (January 2019), https://doi.org/10.1037/hea0000638.

Heehyul Moon et al., "Dementia Prevalence in Older Adults: Variation by Race/Ethnicity and Immigrant Status," *American Journal of Geriatric Psychiatry* 27, no. 3 (March 2019), https://doi.org/10.1016/j.jagp.2018.11.003.

Patricia Coogan et al., "Experiences of Racism and Subjective Cognitive Function in African American Women," *Alzheimer's & Dementia: Diagnosis, Assessment & Disease Monitoring* 12, no. 1 (July 2020), https://doi.org/10.1002%2Fdad2.12067.

135. Kristine Yaffe et al., "Posttraumatic Stress Disorder and Risk of Dementia among US Veterans," *Archives of General Psychiatry* 67, no. 6 (June 2010), https://doi.org/10.1001/archgenpsychiatry.2010.61.

Salah U. Qureshi et al., "Greater Prevalence and Incidence of Dementia in Older Veterans with Posttraumatic Stress Disorder," *Journal of the American Geriatrics Society* 58, no. 9 (September 2010), https://doi.org/10.1111/j.1532-5415.2010.02977.x.

Sean A. P. Clouston et al., "Cognitive Impairment among World Trade Center Responders: Long-Term Implications of Re-experiencing the 9/11 Terrorist Attacks," *Alzheimer's & Dementia: Diagnosis, Assessment & Disease Monitoring* 4 (August 2016), https://doi.org/10.1016/j.dadm.2016.08.001.

136. Robert T. Carter, "Racism and Psychological and Emotional Injury: Recognizing and Assessing Race-Based Traumatic Stress," *Counseling Psychologist* 35, no. 1 (2007), https://doi.org/10.1177/0011000006292033.

Robert T. Carter and Jessica Forsyth, "Reactions to Racial Discrimination: Emotional Stress and Help-Seeking Behaviors," *Psychological Trauma* 2, no. 3 (2010), https://psycnet.apa.org/doi/10.1037/a0020102.

Hugh F. Butts, "The Black Mask of Humanity: Racial/Ethnic Discrimination and Post-Traumatic Stress Disorder," *Journal of the American Academy of Psychiatry and the Law Online* 30, no. 3 (2002), https://jaapl.org/content/30/3/336.long.

Chalsa M. Loo, John A. Fairbank, and Claude M. Chemtob, "Adverse Race-Related Events as A Risk Factor for Posttraumatic Stress Disorder in Asian American Vietnam Veterans," *Journal of Nervous and Mental Disease* 193, no. 7 (July 2005), https://doi.org/10.1097/01.nmd.0000168239.51714.e6.

Katherine Kirkinis et al., "Racism, Racial Discrimination, and Trauma: A Systematic Review of the Social Science Literature," *Ethnicity & Health* 26, no. 3 (April 2021), https://doi.org/10.1080/13557858.2018.1514453.

137. Darnella Frazier, Instagram post via @darnella_frazier03, 2021, https://www.instagram.com/p/CPT5_oIBlie/.

138. "Teen Who Filmed George Floyd's Murder Given Journalism Award," BBC, June 11, 2021, https://www.bbc.com/news/world-us-canada-57449229.

139. Felicia Campbell and Pamela Valera, "'The Only Thing New Is the Cameras': A Study of US College Students' Perceptions of Police Violence on Social Media," *Journal of Black Studies* 51, no. 7 (2020), https://doi.org/10.1177/07435584221144975.
140. Andrea Burri et al., "Childhood Trauma and PTSD Symptoms Increase the Risk of Cognitive Impairment in a Sample of Former Indentured Child Laborers in Old Age," *PLoS One* 8, no. 2 (2013), https://doi.org/10.1371/journal.pone.0057826.
Lianne Hoeijmakers et al., "A Preclinical Perspective on the Enhanced Vulnerability to Alzheimer's Disease After Early-Life Stress," *Neurobiology of Stress* 8, (February 2018), https://doi.org/10.1016%2Fj.ynstr.2018.02.003.
Kylie Radford et al., "Childhood Stress and Adversity Is Associated with Late-Life Dementia in Aboriginal Australians," *American Journal of Geriatric Psychiatry* 25, no. 10 (October 2017), https://doi.org/10.1016/j.jagp.2017.05.008.
Kylie Radford et al., "Factors Associated with the High Prevalence of Dementia in Older Aboriginal Australians," *Journal of Alzheimer's Disease* 70, suppl. 1 (2019), https://doi.org/10.3233/jad-180573.
141. Barbara A. Caldwell and Nancy Redeker, "Sleep and Trauma: An Overview," *Issues in Mental Health Nursing* 26, no. 7 (August/September 2005), https://doi.org/10.1080/01612840591008294.
142. David S. Curtis et al., "Habitual Sleep as a Contributor to Racial Differences in Cardiometabolic Risk," *Proceedings of the National Academy of Sciences* 114, no. 33 (August 2017), https://doi.org/10.1073/pnas.1618167114.
Girardin Jean-Louis et al., "Differential Increase in Prevalence Estimates of Inadequate Sleep Among Black and White Americans," *BMC Public Health* 15 (November 2015), https://doi.org/10.1186/s12889-015-2500-0.
Andrew S. Tubbs et al., "Racial/Ethnic Minorities have Greater Declines in Sleep Duration with Higher Risk of Cardiometabolic Disease: An Analysis of the US National Health Interview Survey," *Sleep Epidemiology* 2 (December 2022), https://doi.org/10.1016/j.sleepe.2022.100022.
Kenneth Anujuo et al., "Ethnic Differences in Self-Reported Sleep Duration in the Netherlands—The Helius Study," *Sleep Medicine* 15, no. 9 (September 2014), https://doi.org/10.1016/j.sleep.2014.04.019.
Kenneth O. Anujuo et al., "Ethnic differences in Sleep Duration at 5 Years, and its Relationship with Overweight and Blood Pressure," *European Journal of Public Health* 26, no. 6 (December 2016), https://doi.org/10.1093/eurpub/ckw084.
143. Nate Seltenrich, "Inequality of Noise Exposures: A Portrait of the United States," *Environmental Health Perspectives* 125, no. 9 (September 2017), https://doi.org/10.1289/EHP2471.
Shawna M. Nadybal, Timothy W. Collins, and Sara E. Grineski, "Light Pollution Inequities in the Continental United States: A Distributive Environmental Justice Analysis," *Environmental Research* 189 (October 2020), https://doi.org/10.1016/j.envres.2020.109959.
Bongki Woo et al., "Residential Segregation and Racial/Ethnic Disparities in Ambient Air Pollution," *Race and Social Problems* 11, no. 1 (March 2019), https://doi.org/10.1007/s12552-018-9254-0.
"A Demographic Profile of US Workers Around the Clock," Population Reference Bureau, September 18, 2008, https://www.prb.org/resources/a-demographic-profile-of-u-s-workers-around-the-clock/.
Night Shift Work, IARC Monographs, vol. 124 (Lyon: World Health Organization, 2020), https://www.iarc.who.int/news-events/iarc-monographs-volume-124-night-shift-work/.
144. Amy Lehrner and Rachel Yehuda, "Cultural Trauma and Epigenetic Inheritance," *Development and Psychopathology* 30, no. 5 (December 2018), https://doi.org/10.1017/s0954579418001153.

Zahava Solomon, Moshe Kotler, and Mario Mikulincer, "Combat-Related Posttraumatic Stress Disorder among Second-Generation Holocaust Survivors: Preliminary Findings," *American Journal of Psychiatry* 145, no. 7 (July 1988), https://doi.org/10.1176/ajp.145.7.865.

Rachel Yehuda et al., "Relationship between Posttraumatic Stress Disorder Characteristics of Holocaust Survivors and their Adult Offspring," *American Journal of Psychiatry* 155, no. 6 (June 1998), https://doi.org/10.1176/ajp.155.6.841.

Rachel Yehuda et al., "Vulnerability to Posttraumatic Stress Disorder in Adult Offspring of Holocaust Survivors," *American Journal of Psychiatry* 155, no. 9 (September 1998), https://doi.org/10.1176/ajp.155.9.1163.

Rachel Yehuda, Sarah L. Halligan, and Linda M. Bierer, "Relationship of Parental Trauma Exposure and PTSD to PTSD, Depressive and Anxiety Disorders in Offspring," *Journal of Psychiatric Research* 35, no. 5 (September/October 2001), https://doi.org/10.1016/s0022-3956(01)00032-2.

Emma A. Payne and David Berle, "Posttraumatic Stress Disorder Symptoms among Offspring of Holocaust Survivors: A Systematic Review and Meta-Analysis," *Traumatology* 27, no. 3 (2021), https://psycnet.apa.org/doi/10.1037/trm0000269.

145. John Cloud, "Why Your DNA Isn't Your Destiny," *Time*, January 6, 2010, https://content.time.com/time/magazine/article/0,9171,1952313,00.html.

Undraga Schagdarsurengin and Klaus Steger, "Epigenetics in Male Reproduction: Effect of Paternal Diet on Sperm Quality and Offspring Health," *Nature Reviews Urology* 13, no. 10 (October 2016), https://doi.org/10.1038/nrurol.2016.157.

L. H. Lumey et al., "Cohort Profile: the Dutch Hunger Winter Families Study," *International Journal of Epidemiology* 36, no. 6 (January 2008), https://doi.org/10.1093/ije/dym126.

Bastiaan T. Heijmans et al., "Persistent Epigenetic Differences Associated with Prenatal Exposure to Famine in Humans," *Proceedings of the National Academy of Sciences* 105, no. 44 (November 2008), https://doi.org/10.1073/pnas.0806560105.

Lars Olov Bygren, Gunnar Kaati, and Sören Edvinsson, "Longevity Determined by Paternal Ancestors' Nutrition During Their Slow Growth Period," *Acta Biotheoretica* 49 (2001), https://doi.org/10.1023/a:1010241825519.

Gunnar Kaati, Lars Olov Bygren, and Sören Edvinsson, "Cardiovascular and Diabetes Mortality Determined by Nutrition During Parents' and Grandparents' Slow Growth Period," *European Journal of Human Genetics* 10, no. 11 (November 2002), https://doi.org/10.1038/sj.ejhg.5200859.

Marcus E. Pembrey et al., "Sex-Specific, Male-Line Transgenerational Responses in Humans," *European Journal of Human Genetics* 14 (2006), https://doi.org/10.1038/sj.ejhg.5201538.

Gunnar Kaati et al., "Transgenerational Response to Nutrition, Early Life Circumstances and Longevity," *European Journal of Human Genetics* 15 (2007), https://doi.org/10.1038/sj.ejhg.5201832.

146. Gerda Egger et al., "Epigenetics in Human Disease and Prospects for Epigenetic Therapy," *Nature* 429, no. 6990 (May 2004), https://doi.org/10.1038/nature02625.

Christina M. Sheerin et al., "The Genetics and Epigenetics of PTSD: Overview, Recent Advances, and Future Directions," *Current Opinion in Psychology* 14, (April 2017), https://doi.org/10.1016/j.copsyc.2016.09.003.

147. Rachel Yehuda et al., "Holocaust Exposure Induced Intergenerational Effects on FKBP5 Methylation," *Biological Psychiatry* 80, no. 5 (September 2016), https://doi.org/10.1016/j.biopsych.2015.08.005.

148. Joy DeGruy, *Post Traumatic Slave Syndrome: America's Legacy of Enduring Injury and Healing* (Portland, OR: Joy DeGruy Publications, 2005).

149. Rawia Liverpool, "My Story with Self-Esteem," Recipes4Change, February 3, 2018, https://www.recipes4change.com/blog/2018/10/18/my-story-with-self-esteem.

150. Alice Walker, *In Search of Our Mothers' Gardens* (San Diego: Harcourt Brace Jovanovich, 1982).
Linda M. Burton et al., "Critical Race Theories, Colorism, and the Decade's Research on Families of Color," *Journal of Marriage and Family* 72, no. 3 (June 2010), https://doi.org/10.1111/j.1741-3737.2010.00712.x.
151. Andrew Leung, "If You Google 'Unprofessional Hairstyles for Work,' These Are the Problematic Results," Mic, April 7, 2016, https://www.mic.com/articles/140092/if-you-google-unprofessional-hairstyles-for-work-these-are-the-problematic-results.
Eleanor Busby, "Pupil Repeatedly Sent Home from School Over Afro Hair Wins £8,500 Payout," *Independent*, February 7, 2020, https://www.independent.co.uk/news/education/education-news/afro-hair-discrimation-student-legal-action-payout-ruby-williams-urswick-school-a9323466.html.
Charley Locke, "6 Kids Speak Out Against Hair Discrimination," *New York Times Magazine*, April 22, 2022, https://www.nytimes.com/2022/04/22/magazine/kids-hair-discrimination.html.
Greg Nicholson, "South African Students Speak Out Against 'Aggressive' Ban on Afro Hair," *Guardian*, August 31, 2016, https://www.theguardian.com/world/2016/aug/31/south-african-students-speak-out-ban-afro-hair-pretoria-school.
Chanté Griffin, "How Natural Black Hair at Work Became a Civil Rights Issue," JSTOR Daily, July 3, 2019, https://daily.jstor.org/how-natural-black-hair-at-work-became-a-civil-rights-issue/.
Nadine White, "Black Hair Discrimination Must be Banned, Equalities Watchdog Told," *Independent*, October 20, 2021, https://www.independent.co.uk/news/uk/home-news/black-hair-discrimination-watchdog-equalities-b1941567.html.
152. Patricia F. Coogan et al., "Hair Product Use and Breast Cancer Incidence in The Black Women's Health Study," *Carcinogenesis* 42, no. 7 (July 2021), https://doi.org/10.1093/carcin/bgab041.
153. Louise A. Brinton et al., "Skin Lighteners and Hair Relaxers as Risk Factors for Breast Cancer: Results from The Ghana Breast Health Study," *Carcinogenesis* 39, no. 4 (April 2018), https://doi.org/10.1093%2Fcarcin%2Fbgy002.
154. Tamarra James-Todd et al., "Childhood Hair Product Use and Earlier Age at Menarche in a Racially Diverse Study Population: A Pilot Study," *Annals of Epidemiology* 21, no. 6 (June 2011), https://doi.org/10.1016/j.annepidem.2011.01.009.
Tamarra James-Todd et al., "Hormonal Activity in Commonly Used Black Hair Care Products: Evaluating Hormone Disruption as a Plausible Contribution to Health Disparities," *Journal of Exposure Science & Environmental Epidemiology* 31, no. 3 (May 2021), https://doi.org/10.1038/s41370-021-00335-3.
155. Tamara Gilkes Borr, "The Hidden Cost of Black Hair," *The Economist*, YouTube, 2021, https://www.youtube.com/watch?v=xgbdOn6uva8.
156. "Cosmetics: Scientific and Technical Assessment," European Commission, https://single-market-economy.ec.europa.eu/sectors/cosmetics/scientific-and-technical-assessment_en.
"Small Businesses & Homemade Cosmetics: Fact Sheet," US Food and Drug Administration, https://www.fda.gov/cosmetics/resources-industry-cosmetics/small-businesses-homemade-cosmetics-fact-sheet.
"EDC-Free Europe Campaign Urges EU to Ban the Use of Endocrine Disruptors in Cosmetics," EDC-Free Europe, September 16, 2021, https://www.edc-free-europe.org/articles/letters/edc-free-europe-campaign-urges-eu-to-ban-the-use-of-endocrine-disruptors.
157. T. Joel Wade, Melanie Judkins Romano, and Leslie Blue, "The Effect of African American Skin Color on Hiring Preferences," *Journal of Applied Social Psychology* 34, no. 12 (December 2004), https://doi.org/10.1111/J.1559-1816.2004.TB01991.X.
Matthew S. Harrison and Kecia M. Thomas, "The Hidden Prejudice in Selection: A Research Investigation on Skin Color Bias," *Journal of Applied Social Psychology* 39, no. 1 (2009), https://doi.org/10.1111/j.1559-1816.2008.00433.x.

Gianna Toboni, "Why India's Fair Skin Business Is Booming," VICE, YouTube, 2020, https://www.youtube.com/watch?v=BlOHSbf9XGI.

Itisha Nagar, "The Unfair Selection: A Study on Skin-Color Bias in Arranged Indian Marriages," *Sage Open* 8, no. 2 (2018), https://doi.org/10.1177/2158244018773149.

Meera Senthilingam, Pallabi Munsi, and Vanessa Offiong, "Skin Whitening: What Is It, What Are the Risks and Who Profits?," CNN, January 25, 2022, https://edition.cnn.com/2022/01/25/world/as-equals-skin-whitening-global-market-explainer-intl-cmd/index.html.

158. "Mercury in Skin Lightening Products," World Health Organization, November 3, 2019, https://www.who.int/publications/i/item/WHO-CED-PHE-EPE-19.13.

159. *Toxic Expose: Online Trade of Mercury-Containing Skin Whitening Cosmetics in the Philippines* (Ecological Waster Coalition of the Philippines, 2021), https://ipen.org/documents/toxic-expose-online-trade-mercury-containing-skin-whitening-cosmetics-philippines.

160. Karl Peltzer, Supa Pengpid, and Caryl James, "The Globalization of Whitening: Prevalence of Skin Lighteners (or Bleachers) Use and Its Social Correlates Among University Students in 26 Countries," *International Journal of Dermatology* 55, no. 2 (February 2016), https://doi.org/10.1111/ijd.12860.

161. Joanne Laxamana Rondilla, "Colonial Faces: Beauty and Skin Color Hierarchy in the Philippines and the US" (PhD diss., University of California, Berkeley, 2012), https://escholarship.org/uc/item/9523k0nb.

162. Nadra Kareem Nittle, "The Roots of Colorism, or Skin Tone Discrimination," ThoughtCo, February 28, 2021, https://www.thoughtco.com/what-is-colorism-2834952.

"Colourism In India—A History," Dark Is Beautiful, https://www.darkisbeautiful.in/colourism-in-india/.

Suresh Jungari and Priyanka Bomble, "Caste-Based Social Exclusion and Health Deprivation in India," *Journal of Exclusion Studies* 3, no. 2 (January 2013), http://dx.doi.org/10.5958/j.2231-4555.3.2.011.

163. Raksha Thapa et al., "Caste Exclusion and Health Discrimination in South Asia: A Systematic Review," *Asia Pacific Journal of Public Health* 33, no. 8 (November 2021), https://doi.org/10.1177/10105395211014648.

Suresh Jungari, Baby Sharma, and Dhananjay Wagh, "Beyond Maternal Mortality: A Systematic Review of Evidences on Mistreatment and Disrespect During Childbirth in Health Facilities in India," *Trauma Violence & Abuse* 22, no. 4 (October 2021), https://doi.org/10.1177/1524838019881719.

164. Megha Rajagopalan, "'Black Lives Matter,' Say These Companies That Sell Skin Lightening Products," BuzzFeed News, June 14, 2020, https://www.buzzfeed.com/meghara/skin-lightening-cream-black-lives-matter-companies.

"Unilever Evolves Skin Care Portfolio to Embrace a More Inclusive Vision of Beauty," Unilever, June 25, 2020, https://www.unilever.com/news/press-and-media/press-releases/2020/unilever-evolves-skin-care-portfolio-to-embrace-a-more-inclusive-vision-of-beauty/.

Martinne Geller, "L'Oreal to Drop Words such as 'Whitening' from Skin Products," Reuters, June 27, 2020, https://www.reuters.com/article/us-l-oreal-whitening-idCAKBN23X224.

Georgina Caldwell, "By Any Other Name? P&G to Re-Brand Skin Lightening Creams," Global Cosmetics News, September 28, 2020, https://www.globalcosmeticsnews.com/by-any-other-name-pg-to-re-brand-skin-lightening-creams/.

Martinne Geller, "Johnson & Johnson Drops Skin-Whitening Creams," Reuters, June 20, 2020, https://www.reuters.com/article/us-johnson-johnson-whitening-idUSKBN23Q2BZ.

Prim Chuwirich, Malavika Kaur Makol, and Ragini Saxena, "Skin-Whitening Products Are Still Big Business in Asia," Bloomberg, September 23, 2021, https://www.bloomberg.com/news/articles/2021-09-22/skin-whitening-creams-remain-big-business-in-asia-despite-purge.

165. Darlene Diep, "Stevens-Johnson Syndrome in a Patient of Color: A Case Report and an Assessment of Diversity in Medical Education Resources," *Cureus* 14, no. 2 (February 2022), https://doi.org/10.7759/cureus.22245.

166. Patricia Louie and Rima Wilkes, "Representations of Race and Skin Tone in Medical Textbook Imagery," *Social Science & Medicine* 202 (April 2018), https://doi.org/10.1016/j.socscimed.2018.02.023.

Neil Singh, "Decolonising Dermatology: Why Black and Brown Skin Need Better Treatment," *Guardian*, August 13, 2020, https://www.theguardian.com/society/2020/aug/13/decolonising-dermatology-why-black-and-brown-skin-need-better-treatment.

167. Jonathan Dutt et al., "Stevens-Johnson Syndrome: A Perplexing Diagnosis," *Cureus* 12, no. 3 (March 2020), https://doi.org/10.7759/cureus.7374.

Danilo Buonsenso et al., "Impact of Diversity in Training Resources on Self-Confidence in Diagnosing Skin Conditions Across a Range of Skin Tones: An International Survey," *Frontiers in Pediatrics* 10 (February 2022), https://doi.org/10.3389/fped.2022.837552.

168. Sean M. Dawes et al., "Racial Disparities in Melanoma Survival," *Journal of the American Academy of Dermatology* 75, 983–991 (2016), https://doi.org/10.1016/j.jaad.2016.06.006.

169. Christina M. Correnti, "Racial Disparities in Fifth-Grade Sun Protection: Evidence from the Healthy Passages Study," *Pediatric Dermatology* 35, no. 5 (September 2018), https://doi.org/10.1111/pde.13550.

Sally Wadyka, "Sun Safety Tips for the Entire Family," *Consumer Reports*, May 20, 2020, https://www.consumerreports.org/sun-protection/sun-safety-tips-for-the-entire-family-a7948562866/.

Amel Ben Allal, "Darker Skin Tones Don't Need Sunscreen: Fact or Fiction?," 4.5.6 Skin, https://456skin.com/blogs/4-5-6-talks/darker-skin-tones-don-t-need-sunscreen-fact-or-fiction.

170. Sharon Belmo, "Viewpoint: Why Dermatology for Skin of Colour Matters," GPonline, March 15, 2021, https://www.gponline.com/viewpoint-why-dermatology-skin-colour-matters/dermatology/article/1709994.

171. J. C. Lester, S. C. Taylor, and M.-M. Chren, "Under-Representation of Skin of Colour in Dermatology Images: Not Just an Educational Issue," *British Journal of Dermatology* 180, no. 6 (June 2019), https://doi.org/10.1111/bjd.17608.

172. Richard L. Street Jr. et al., "How Does Communication Heal? Pathways Linking Clinician–Patient Communication to Health Outcomes," *Patient Education and Counseling* 74, no. 3 (March 2009), https://doi.org/10.1016/j.pec.2008.11.015.

173. John Eligon, "Black Doctor Dies of Covid-19 After Complaining of Racist Treatment," *New York Times*, December 23, 2020, https://www.nytimes.com/2020/12/23/us/susan-moore-black-doctor-indiana.html.

Bill Hutchinson, "Black Doctor Dies of COVID After Racist Treatment Complaints," ABC News, December 25, 2020, https://abcnews.go.com/US/black-doctor-dies-covid-alleging-hospital-mistreatment-black/story?id=74878119.

174. Aletha Maybank et al., "Say Her Name: Dr. Susan Moore," *Opinion, Washington Post*, December 26, 2020, https://www.washingtonpost.com/opinions/2020/12/26/say-her-name-dr-susan-moore/.

175. Paulyne Lee et al., "Racial and Ethnic Disparities in the Management of Acute Pain in US Emergency Departments: Meta-Analysis and Systematic Review," *American Journal of Emergency Medicine* 37, no. 9 (September 2019), https://doi.org/10.1016/j.ajem.2019.06.014.

Katarina Rukavina et al., "Ethnic Disparities in Treatment of Chronic Pain in Individuals with Parkinson's Disease Living in the United Kingdom," *Movement Disorders Clinical Practice* 9, no. 3 (April 2022), https://doi.org/10.1002%2Fmdc3.13430.

176. Linda Villarosa, *Under the Skin: The Hidden Toll of Racism on American Lives and on the Health of Our Nation* (New York: Doubleday, 2022).

177. Philip P. Goodney et al., *Variation in the Care of Surgical Conditions: Diabetes and Peripheral Arterial Disease* (Dartmouth Institute for Health Policy & Clinical Practice, 2014), https://data.dartmouthatlas.org/downloads/reports/Diabetes_report_10_14_14.pdf.
Lizzie Presser, "The Black American Amputation Epidemic," ProPublica, May 19, 2020, https://features.propublica.org/diabetes-amputations/black-american-amputation-epidemic/.
Tyler S. Durazzo, Stanley Frencher, and Richard Gusberg, "Influence of Race on the Management of Lower Extremity Ischemia: Revascularization vs Amputation," *JAMA Surg* 148, no. 7 (2013), https://doi.org/10.1001/jamasurg.2013.1436.
Shipra Arya et al., "Race and Socioeconomic Status Independently Affect Risk of Major Amputation in Peripheral Artery Disease," *Journal of the American Heart Association* 7, no. 2 (January 2018), https://doi.org/10.1161/jaha.117.007425.
Meghan B. Brennan et al., "Association of Race, Ethnicity, and Rurality with Major Leg Amputation or Death Among Medicare Beneficiaries Hospitalized with Diabetic Foot Ulcers," *JAMA Network Open* 5, no. 4 (April 2022), https://doi.org/10.1001%2Fjamanetworkopen.2022.8399.
Richard M. Mizelle Jr., "Diabetes, Race, and Amputations," *Lancet* 397, no. 10281 (April 2021), https://doi.org/10.1016/s0140-6736(21)00724-8.
178. "Medical Workforce Race Equality Standard 2020 Data Report," NHS, July 2021, https://www.england.nhs.uk/publication/medical-workforce-race-equality-standard-2020-data-report/.
179. Brad N. Greenwood et al., "Physician–Patient Racial Concordance and Disparities in Birthing Mortality for Newborns," *Proceedings of the National Academy of Sciences* 117, no. 35 (September 2020), https://doi.org/10.1073/pnas.1913405117.
Marcella Alsan, Owen Garrick, and Grant Graziani, "Does Diversity Matter for Health? Experimental Evidence from Oakland," *American Economic Review* 109, no. 12 (December 2019), https://doi.org/10.1257/aer.20181446.
180. "Former NFL Players Speak Out about Race-Based Concussion Damage Assessment," Nightline/ABC News, YouTube, 2021, https://www.youtube.com/watch?v=j7daTuaByy4.
Jesse Mez et al., "Clinicopathological Evaluation of Chronic Traumatic Encephalopathy in Players of American Football," *Journal of the American Medical Association* 318, no. 4 (July 2017), https://doi.org/10.1001/jama.2017.8334.
Michael L. Alosco et al., "Association of White Matter Rarefaction, Arteriolosclerosis, and Tau with Dementia in Chronic Traumatic Encephalopathy," *JAMA Neurology* 76, no. 11 (November 2019), https://doi.org/10.1001/jamaneurol.2019.2244.
181. "NFL Sued for Limiting Black Players' Access to Concussion Benefits," Associated Press, August 25, 2020, https://apnews.com/PR%20Newswire/916620aa21015862bbfa6051b620b7b9.
Maryclaire Dale, "Judge Tosses Suit Over 'Race-Norming' in NFL Dementia Tests," Associated Press, March 8, 2021, https://apnews.com/article/judge-tosses-race-norming-suit-nfl-dementia-tests-ed61e9917fcd94d79c1fe79a13e037e0.
Maryclaire Dale, "NFL Pledges to Halt 'Race-Norming,' Review Black Claims," Associated Press, June 3, 2021, https://apnews.com/article/pa-state-wire-race-and-ethnicity-health-nfl-sports-205b304c0c3724532d74fc54e58b4d1d.
182. Layal Liverpool, "How Medical Tests Have Built-In Discrimination Against Black People," *New Scientist*, July 14, 2021, https://www.newscientist.com/article/mg25133434-100-how-medical-tests-have-built-in-discrimination-against-black-people/.
183. Fabiola Cineas, "'Race Norming' and the Long Legacy of Medical Racism, Explained," Vox, July 8, 2021, https://www.vox.com/22528334/race-norming-medical-racism.
Robert K. Heaton, *Revised Comprehensive Norms for an Expanded Halstead-Reitan Battery: Demographically Adjusted Neuropsychological Norms for African American and Caucasian Adults*, professional manual (Psychological Assessment Resources, 2004), cited in Philip G.

Gasquoine, "Race-Norming of Neuropsychological Tests," *Neuropsychology Review* 19, no. 2 (June 2009), https://doi.org/10.1007/s11065-009-9090-5.

Gasquoine, "Race-Norming of Neuropsychological Tests."

Kaitlin B. Casaletto et al., "Demographically Corrected Normative Standards for the English Version of the NIH Toolbox Cognition Battery," *Journal of the International Neuropsychological Society* 21, no. 5 (May 2015), https://doi.org/10.1017/s1355617715000351.

184. Katherine L. Possin, Elena Tsoy, and Charles C. Windon, "Perils of Race-Based Norms in Cognitive Testing: The Case of Former NFL Players," *JAMA Neurology* 78, no. 4 (April 2021), https://doi.org/10.1001%2Fjamaneurol.2020.4763.

185. Dorothy Roberts, "The Problem with Race-Based Medicine," TEDMED talk, 2015, https://www.tedmed.com/talks/show?id=530900.

186. Vishal Duggal et al., "National Estimates of CKD Prevalence and Potential Impact of Estimating Glomerular Filtration Rate without Race," *Journal of the American Society of Nephrology* 32, no. 6 (June 2021), https://doi.org/10.1681/asn.2020121780.

187. Vanessa Grubbs, "Precision in GFR Reporting: Let's Stop Playing the Race Card," *Clinical Journal of the American Society of Nephrology* 15, no. 8 (August 2020), https://doi.org/10.2215/cjn.00690120.

Nwamaka Denise Eneanya, Wei Yang, and Peter Philip Reese, "Reconsidering the Consequences of Using Race to Estimate Kidney Function," *Journal of the American Medical Association* 322, no. 2 (July 2019), https://doi.org/10.1001/jama.2019.5774.

188. Andrew S. Levey et al., "A More Accurate Method to Estimate Glomerular Filtration Rate from Serum Creatinine: A New Prediction Equation. Modification of Diet in Renal Disease Study Group," *Annals of Internal Medicine* 130, no. 6 (March 1999), https://doi.org/10.7326/0003-4819-130-6-199903160-00002.

189. Andrew S. Levey et al., "A New Equation to Estimate Glomerular Filtration Rate," *Annals of Internal Medicine* 150, no. 9 (May 2009), https://doi.org/10.7326/0003-4819-150-9-200905050-00006.

190. Jennifer Tsai, "Jordan Crowley Would Be in Line for a Kidney—if He Were Deemed White Enough," *Slate*, June 27, 2021, https://slate.com/technology/2021/06/kidney-transplant-dialysis-race-adjustment.html.

Jennifer W. Tsai et al., "Evaluating the Impact and Rationale of Race-Specific Estimations of Kidney Function: Estimations from US NHANES, 2015–2018" *EClinicalMedicine* (November 2021), https://doi.org/10.1016/j.eclinm.2021.101197.

191. *Jordan Crowley* v. *Strong Memorial Hospital of the University of Rochester; Kaleida Health and UBMD Physicians Group*, [2021] New York Western District.

192. *KDIGO 2012 Clinical Practice Guideline for the Evaluation and Management of Chronic Kidney Disease* (KDIGO, 2013).

193. Naomi T. Nkinsi and Bessie A. Young, "How the University of Washington Implemented a Change in eGFR Reporting," *Kidney360* 3, no. 3 (March 2022), https://doi.org/10.34067%2FKID.0006522021.

Dorothy E. Roberts, "Abolish Race Correction," *Lancet* 397, no. 10268 (January 2021), https://doi.org/10.1016/s0140-6736(20)32716-1.

194. "Chronic Kidney Disease in Adults: Assessment and Management," National Institute for Health and Care Excellence, July 23, 2014, https://www.nice.org.uk/guidance/cg182.

"Guideline Chronic Kidney Disease: Draft for Consultation, January 2021," National Institute for Health and Care Excellence, January 2021, https://www.nice.org.uk/guidance/ng203/documents/draft-guideline.

195. Rouvick M. Gama et al., "Estimated Glomerular Filtration Rate Equations in People of Self-Reported Black Ethnicity in the United Kingdom: Inappropriate Adjustment for Ethnicity May Lead to Reduced Access to Care," *PLoS One* 16, no. 8 (August 2021), https://doi.org/10.1371/journal.pone.0255869.

"Chronic Kidney Disease: Assessment and Management," National Institute for Health and Care Excellence, August 25, 2021, https://www.nice.org.uk/guidance/ng203.

196. Layal Liverpool, "Kidney Test Adjustment Based on Ethnicity Cut from UK Medical Guidance," *New Scientist*, August 25, 2021, https://www.newscientist.com/article/2288008-kidney-test-adjustment-based-on-ethnicity-cut-from-uk-medical-guidance/.
197. "NKF and ASN Release New Way to Diagnose Kidney Diseases," National Kidney Foundation, September 23, 2021, https://www.kidney.org/news/nkf-and-asn-release-new-way-to-diagnose-kidney-diseases.
198. "OPTN Board Approves Elimination of Race-Based Calculation for Transplant Candidate Listing," Organ Procurement & Transplantation Network, June 28, 2022, https://optn.transplant.hrsa.gov/news/optn-board-approves-elimination-of-race-based-calculation-for-transplant-candidate-listing/.
199. Heidi L. Lujan and Stephen E. DiCarlo, "Science Reflects History as Society Influences Science: Brief History of 'Race,' 'Race Correction,' and the Spirometer,"' *Advances in Physiology Education* 42, no. 1 (June 2018), https://doi.org/10.1152/advan.00196.2017.
200. A. T. Moffett et al., "The Impact of Race Correction on the Interpretation of Pulmonary Function Testing Among Black Patients," in *Impact of Race, Ethnicity, and Social Determinants on Individuals with Lung Diseases* (American Thoracic Society, 2021), https://doi.org/10.1164/ajrccm-conference.2021.203.1_MeetingAbstracts.A1030.
201. Brian L. Graham et al., "Standardization of Spirometry 2019 Update. An Official American Thoracic Society and European Respiratory Society Technical Statement," *American Journal of Respiratory and Critical Care Medicine* 200, no. 8 (October 2019), https://doi.org/10.1164/rccm.201908-1590st.
202. "ATS Publishes Official Statement on Race, Ethnicity and Pulmonary Function Test Interpretation," American Thoracic Society, 2023, https://www.thoracic.org/about/newsroom/press-releases/journal/2023/pft-and-race-official-statement.php.
203. Richard S. Garcia, "The Misuse of Race in Medical Diagnosis," *Pediatrics* 113, no. 5 (May 2004), https://doi.org/10.1542/peds.113.5.1394.
204. Michael Boyle, "Addressing Racism and Discrimination," Cystic Fibrosis Foundation, July 8, 2020, https://www.cff.org/community-posts/2020-07/addressing-racism-and-discrimination.

    Rita Rubin, "Tackling the Misconception That Cystic Fibrosis Is a 'White People's Disease,'" *Journal of the American Medical Association* 325, no. 23 (June 2021), https://doi.org/10.1001/jama.2021.5086.

    Alexandra Power-Hays and Patrick T. McGann, "When Actions Speak Louder Than Words—Racism and Sickle Cell Disease," *New England Journal of Medicine* 383, no. 20 (November 2020), https://doi.org/10.1056/nejmp2022125.

    Noone's Listening: An Inquiry into the Avoidable Deaths and Failures of Care for Sickle Cell Patients in Secondary Care (Sickle Cell Society, 2021), https://www.sicklecellsociety.org/no-ones-listening/.
205. Frédéric B. Piel et al., "Global Distribution of the Sickle Cell Gene and Geographical Confirmation of the Malaria Hypothesis," *Nature Communications* 1 (November 2010), https://doi.org/10.1038/ncomms1104.

    Michael Aidoo et al., "Protective Effects of the Sickle Cell Gene Against Malaria Morbidity and Mortality," *Lancet* 359, no. 9314 (April 2002), https://doi.org/10.1016/s0140-6736(02)08273-9.
206. Darshali A. Vyas, Leo G. Eisenstein, and David S. Jones, "Hidden in Plain Sight—Reconsidering the Use of Race Correction in Clinical Algorithms," *New England Journal of Medicine* 383, no. 9 (August 2020), https://doi.org/10.1056/NEJMms2004740.

    Elizabeth Warren, "Warren, Wyden, Booker, and Lee Question the Use of Race-Based Algorithms in Standard Medical Practice," September 24, 2020, https://www.warren.

senate.gov/newsroom/press-releases/warren-wyden-booker-and-lee-question-the-use-of-race-based-algorithms-in-standard-medical-practice.

207. "Hypertension in Adults: Diagnosis and Management," National Institute for Health and Care Excellence, August 28, 2019, https://www.nice.org.uk/guidance/ng136.
Dipesh P. Gopal and Rohin Francis, "Does Race Belong in the Hypertension Guidelines?," *Journal of Human Hypertension* 35, no. 10 (October 2021), https://doi.org/10.1038/s41371-020-00414-2.
Dipesh P. Gopal, Grace N. Okoli, and Mala Rao, "Re-thinking the Inclusion of Race in British Hypertension Guidance," *Journal of Human Hypertension* 36, no. 3 (March 2022), https://doi.org/10.1038%2Fs41371-021-00601-9.
208. Dena E. Rifkin et al., "Association of Renin and Aldosterone With Ethnicity and Blood Pressure: The Multi-Ethnic Study of Atherosclerosis," *American Journal of Hypertension* 27, no. 6 (June 2014), https://doi.org/10.1093%2Fajh%2Fhpt276.
209. Layal Liverpool and Jennifer Tsai, "Medicine Must Stop Using Race and Ethnicity to Interpret Test Results," *New Scientist*, November 10, 2021, https://www.newscientist.com/article/mg25233602-900-medicine-must-stop-using-race-and-ethnicity-to-interpret-test-results/.
210. "Fact versus Fiction: Clinical Decision Support Tools and the (Mis)use of Race," House Committee on Ways and Means, 2021, https://democrats-waysandmeans.house.gov/sites/evo-subsites/democrats-waysandmeans.house.gov/files/documents/Fact%20Versus%20Fiction%20Clinical%20Decision%20Support%20Tools%20and%20the%20(Mis)Use%20of%20Race%20(2).pdf.
211. Amanda Shendruk, "Are You Even Trying to Stop Racism if You Don't Collect Data on Race?," *Quartz*, July 8, 2021, https://qz.com/2029525/the-20-countries-that-dont-collect-racial-and-ethnic-census-data.
Carlotta Balestra and Lara Fleischer, "Diversity Statistics in the OECD: How Do OECD Countries Collect Data on Ethnic, Racial and Indigenous Identity?," OECD Statistics Working Papers, 2018, https://doi.org/10.1787/89bae654-en.
Annabelle Timsit, "France's Data Collection Rules Obscure the Racial Disparities of Covid-19," *Quartz*, June 5, 2020, https://qz.com/1864274/france-doesnt-track-how-race-affects-covid-19-outcomes.
212. Joost J. Zwart et al., "Ethnic Disparity in Severe Acute Maternal Morbidity: A Nationwide Cohort Study in the Netherlands," *European Journal of Public Health* 21, no. 2 (April 2011), https://doi.org/10.1093/eurpub/ckq046.
M. Saucedo, C. Deneux Tharaux, and M.-H. Bouvier-Colle, "Understanding Regional Differences in Maternal Mortality: A National Case–Control Study in France," *BJOG* 119, no. 5 (April 2012), https://doi.org/10.1111/j.1471-0528.2011.03220.x.
M. Philibert, C. Deneux Tharaux, and M.-H. Bouvier-Colle, "Can Excess Maternal Mortality among Women of Foreign Nationality be Explained by Suboptimal Obstetric Care?," *BJOG* 115, no. 11 (October 2008), https://doi.org/10.1111/j.1471-0528.2008.01860.x.
M. David et al., "Comparison of Perinatal Data of Immigrant Women of Turkish Origin and German Women—Results of a Prospective Study in Berlin," *Geburtshilfe Frauenheilkd* 74, no. 5 (May 2014), https://doi.org/10.1055%2Fs-0034-1368489.
213. Yin Paradies et al., "Racism as a Determinant of Health: A Systematic Review and Meta-Analysis," *PLoS One* 10, no. 9 (2015), https://doi.org/10.1371/journal.pone.0138511.
214. "A Union of Equality: EU Anti-Racism Action Plan 2020–2025," European Commission, 2020, https://commission.europa.eu/strategy-and-policy/policies/justice-and-fundamental-rights/combatting-discrimination/racism-and-xenophobia/eu-anti-racism-action-plan-2020-2025_en.
"Promotion and Protection of the Human Rights and Fundamental Freedoms of Africans and of People of African Descent Against Excessive Use of Force and Other Human Rights

Violations by Law Enforcement Officers," UN OHCHR, July 9, 2021, https://www.ohchr.org/en/documents/reports/ahrc4753-promotion-and-protection-human-rights-and-fundamental-freedoms-africans.

215. Shaheena Janjuha-Jivraj, "How to Fix Gender Data Bias: Top Tips from Caroline Criado-Perez," *Forbes*, April 8, 2020, https://www.forbes.com/sites/shaheenajanjuhajivrajeurope/2020/04/08/data-bias-is-everyones-problem/.

216. "Public Attitudes to a Covid-19 Vaccine, and their Variations Across Ethnic and Socioeconomic Groups," Royal Society for Public Health, December 2020, https://www.rsph.org.uk/our-work/policy/vaccinations/public-attitudes-to-a-covid-19-vaccine.html.
Elaine Robertson et al., "Predictors of COVID-19 vaccine hesitancy in the UK Household Longitudinal Study," *Brain, Behavior, and Immunity* 94 (May 2021), https://doi.org/10.1016/j.bbi.2021.03.008.
Brian MacKenna et al., "Trends, Regional Variation, and Clinical Characteristics of COVID-19 Vaccine Recipients: A Retrospective Cohort Study in 23.4 Million Patients Using OpenSAFELY," medRxiv, January 26, 2021, https://doi.org/10.1101/2021.01.25.21250356.
Layal Liverpool, "UK Government Won't Say if it has Ethnicity Data for Covid-19 Shots," *New Scientist*, January 12, 2021, https://www.newscientist.com/article/2264611-uk-government-wont-say-if-it-has-ethnicity-data-for-covid-19-shots/.
Layal Liverpool, "We Must Start Publishing Ethnicity Data for Covid-19 Vaccinations," *New Scientist*, January 15, 2021, https://www.newscientist.com/article/2265004-we-must-start-publishing-ethnicity-data-for-covid-19-vaccinations/.
Layal Liverpool, "NHS England Criticised Over Missing Ethnicity Data for Covid-19 Jabs," *New Scientist*, January 27, 2021, https://www.newscientist.com/article/2266017-nhs-england-criticised-over-missing-ethnicity-data-for-covid-19-jabs/
Nick Kituno, "NHS England Agrees to Collect Ethnicity Data Seven Weeks After Covid Vaccinations Began," *Health Service Journal*, January 26, 2021, https://www.hsj.co.uk/primary-care/nhs-england-agrees-to-collect-ethnicity-data-seven-weeks-after-covid-vaccinations-began/7029378.article.
Jordan Kelly-Linden, "NHS to Publish Data on Ethnicity to Help Combat Vaccine Hesitancy," *Telegraph*, January 26, 2021, https://www.telegraph.co.uk/global-health/science-and-disease/nhs-publish-data-ethnicity-help-combat-vaccine-hesitancy/.

217. "Coronavirus and Vaccination Rates in People Aged 70 Years and Over by Socio-Demographic Characteristic, England: 8 December 2020 to 11 March 2021," UK Office for National Statistics, March 29, 2021, https://www.ons.gov.uk/peoplepopulationandcommunity/healthandsocialcare/healthinequalities/bulletins/coronavirusandvaccinationratesinpeopleaged70yearsandoverbysociodemographiccharacteristicengland/8december2020to11march2021.
"A Letter to Loved Ones About the COVID-19 Vaccine—Sir Lenny Henry," NHS, YouTube, 2021, https://www.youtube.com/watch?v=0mKYnTZvIUM.
"Coronavirus (COVID-19) Latest Insights: Vaccines," UK Office for National Statistics, 2022, https://www.ons.gov.uk/peoplepopulationandcommunity/healthandsocialcare/conditionsanddiseases/articles/coronaviruscovid19latestinsights/vaccines.

218. "Guidance Note on the Collection and Use of Equality Data Based on Racial or Ethnic Origin," European Commission, 2021, https://commission.europa.eu/system/files/2022-02/guidance_note_on_the_collection_and_use_of_equality_data_based_on_racial_or_ethnic_origin_final.pdf.
"Updated Compendium of Equality Data Practices," European Union Agency for Fundamental Rights, 2022, https://fra.europa.eu/en/news/2022/updated-compendium-equality-data-practices.
"Equality Data Collection: Facts and Principles," European Network Against Racism, 2016, https://www.enar-eu.org/wp-content/uploads/edc-general_factsheet_final.pdf.

219. Daniel Gyamerah, "11_Glossar Anti-Schwarzer Rassismus_Afrozensus," Each One Teach One, YouTube, 2021, https://www.youtube.com/watch?v=w8Yo_PPwqdQ.
220. Vitor Miranda, "A Resurgence of Black Identity in Brazil? Evidence from an Analysis of Recent Censuses," *Demographic Research* 32 (June 2015), https://dx.doi.org/10.4054/DemRes.2015.32.59.
Lucinda Platt, "Royal Wedding: The UK's Rapidly Changing Mixed-Race Population," BBC, May 15, 2018, https://www.bbc.co.uk/news/uk-44040766.
Silvia Foster-Frau, Ted Mellnick, and Adrian Blanco, "'We're Talking about a Big, Powerful Phenomenon': Multiracial Americans Drive Change," *Washington Post*, October 8, 2021, https://www.washingtonpost.com/nation/2021/10/08/mixed-race-americans-increase-census/.
221. Tina J. Kauh, Jen'nan Ghazal Read, and A. J. Scheitler, "The Critical Role of Racial/Ethnic Data Disaggregation for Health Equity," *Population Research and Policy Review* 40, no. 1 (2021), https://doi.org/10.1007/s11113-020-09631-6.
222. "Why Disaggregate? Disparities in AAPI Health," AAPI Data, 2017, http://aapidata.com/blog/countmein-health-di.
223. Rebecca Nagle, "Native Americans Being Left Out of US Coronavirus Data and Labelled as 'Other,'" *Guardian*, April 24, 2020, https://www.theguardian.com/us-news/2020/apr/24/us-native-americans-left-out-coronavirus-data.
224. Tina Bellon and Nate Raymond, "Bristol-Myers, Sanofi Ordered to Pay Hawaii $834 Million Over Plavix Warning Label," Reuters, February 26, 2021, https://www.reuters.com/business/healthcare-pharmaceuticals/bristol-myers-sanofi-ordered-pay-hawaii-834-mln-over-plavix-warning-label-2021-02-15/.
225. Lauren M. Harnel et al., "Barriers to Clinical Trial Enrolment in Racial and Ethnic Minority Patients with Cancer," *Cancer Control* 23, no. 4 (October 2016), https://doi.org/10.1177/107327481602300404.
226. Jessica L. Mega et al., "Cytochrome P-450 Polymorphisms and Response to Clopidogrel," *New England Journal of Medicine* 360, no. 4 (January 2009), https://doi.org/10.1056/nejmoa0809171.
L. R. Johnston et al., "Suboptimal Response to Clopidogrel and the Effect of Prasugrel in Acute Coronary Syndromes," *International Journal of Cardiology* 167, no. 3 (August 2013), https://doi.org/10.1016/j.ijcard.2012.03.080.
227. Layal Liverpool, "Genomic Medicine Is Deeply Biased Towards White People," *New Scientist*, January 20, 2021, https://www.newscientist.com/article/mg24933180-800-genomic-medicine-is-deeply-biased-towards-white-people/.
Bat-sheva Kerem et al., "Identification of the Cystic Fibrosis Gene: Genetic Analysis," *Science* (1979) 245, no. 4922 (September 1989), https://doi.org/10.1126/science.2570460.
Cystic Fibrosis Genetic Analysis Consortium, "Worldwide Survey of the Delta F508 Mutation—Report from the Cystic Fibrosis Genetic Analysis Consortium," *American Journal of Human Genetics* 47, no. 2 (August 1990), https://www.ncbi.nlm.nih.gov/pmc/articles/PMC1683705/.
Cheryl Stewart and Michael S. Pepper, "Cystic Fibrosis in the African Diaspora," *Annals of the American Thoracic Society* 14, no. 1 (January 2017), https://doi.org/10.1513/annalsats.201606-481fr.
C. Padoa et al., "Cystic Fibrosis Carrier Frequencies in Populations of African Origin," *Journal of Medical Genetics* 36, no. 1 (January 1999), https://doi.org/10.1136/jmg.36.1.41.
228. Amal Jubran and Martin J. Tobin, "Reliability of Pulse Oximetry in Titrating Supplemental Oxygen Therapy in Ventilator-Dependent Patients," *Chest* 97, no. 6 (June 1990), https://doi.org/10.1378/chest.97.6.1420.
Philip E. Bickler, John R. Feiner, and John W. Severinghaus, "Effects of Skin Pigmentation on Pulse Oximeter Accuracy at Low Saturation," *Anesthesiology* 102, no. 4 (April 2005), https://doi.org/10.1097/00000542-200504000-00004.

John R. Feiner, John W. Severinghaus, and Philip E. Bickler, "Dark Skin Decreases the Accuracy of Pulse Oximeters at Low Oxygen Saturation: The Effects of Oximeter Probe Type and Gender," *Anesthesia & Analgesia* 105, suppl. 6 (December 2007), https://doi.org/10.1213/01.ane.0000285988.35174.d9.

Michael W. Sjoding et al., "Racial Bias in Pulse Oximetry Measurement," *New England Journal of Medicine* 383, no. 25 (December 2020), https://doi.org/10.1056/nejmc2029240.

"Pulse Oximetry and Racial Bias: Recommendations for National Healthcare, Regulatory and Research Bodies," NHS Race and Health Observatory, 2021, https://www.nhsrho.org/wp-content/uploads/2021/03/Pulse-oximetry-racial-bias-report.pdf.

229. "Guidance: The Use and Regulation of Pulse Oximeters (Information for Healthcare Professionals)," Medicines and Healthcare Products Regulatory Agency, 2021, https://www.gov.uk/guidance/the-use-and-regulation-of-pulse-oximeters-information-for-healthcare-professionals.

"Government Response to Consultation on the Future Regulation of Medical Devices in the United Kingdom," Medicines and Healthcare products Regulatory Agency, 2022, https://assets.publishing.service.gov.uk/government/uploads/system/uploads/attachment_data/file/1085333/Government_response_to_consultation_on_the_future_regulation_of_medical_devices_in_the_United_Kingdom.pdf.

230. Alicia R. Martin et al., "Clinical Use of Current Polygenic Risk Scores May Exacerbate Health Disparities," *Nature Genetics* 51, no. 4 (April 2019), https://doi.org/10.1038/s41588-019-0379-x.

231. Baris A. Ozdemir et al., "Research Activity and the Association with Mortality," *PLoS One* 10, no. 2 (February 2015), https://doi.org/10.1371/journal.pone.0118253.

Leon Jonker, Stacey Jayne Fisher, and Dave Dagnan, "Patients Admitted to More Research-Active Hospitals Have More Confidence in Staff and Are Better Informed About Their Condition and Medication: Results from a Retrospective Cross-Sectional Study," *Journal of Evaluation in Clinical Practice* 26, no. 1 (February 2020), https://doi.org/10.1111/jep.13118.

Sumit R. Majumdar et al., "Better Outcomes for Patients Treated at Hospitals that Participate in Clinical Trials," *Archives of Internal Medicine* 168, no. 68 (March 2008), https://doi.org/10.1001/archinternmed.2007.124.

232. Will Joice and Andy Tetlow, "Baselines for Improving STEM Participation: Ethnicity STEM Data for Students and Academic Staff in Higher Education 2007/08 to 2018/19," Royal Society, October 15, 2020, https://royalsociety.org/-/media/policy/Publications/2021/trends-ethnic-minorities-stem/Ethnicity-STEM-data-for-students-and-academic-staff-in-higher-education.pdf.

Travis A. Hoppe et al., "Topic Choice Contributes to the Lower Rate of NIH Awards to African-American/Black Scientists," *Science Advances* 5, no. 10 (October 2019), https://doi.org/10.1126/sciadv.aaw7238.

"Detailed Analysis of UKRI Funding Applicants and Awardees Ethnicity, Financial Years 2015–16 to 2019–20," UK Research and Innovation, 2021, https://www.ukri.org/wp-content/uploads/2021/10/UKRI-061021-EthnicityAnalysisReportFinal.pdf.

233. Ziad Obermeyer et al., "Dissecting Racial Bias in an Algorithm Used to Manage the Health of Populations," *Science* 366, no. 6464 (October 2019), https://doi.org/10.1126/science.aax2342.

234. "Accuracy of Claims-Based Risk Scoring Models," Society of Actuaries, 2016, https://www.soa.org/globalassets/assets/Files/Research/research-2016-accuracy-claims-based-risk-scoring-models.pdf.

235. Adam Bohr and Kaveh Memarzadeh, "The Rise of Artificial Intelligence in Healthcare Application," *Artificial Intelligence in Healthcare*, 2020, https://doi.org/10.1016/B978-0-12-818438-7.00002-2.

"Alphabet Is Spending Billions to Become a Force in Health Care," *The Economist*, June 20, 2022, https://www.economist.com/business/2022/06/20/alphabet-is-spending-billions-to-become-a-force-in-health-care.

236. Ruth Hailu, "Fitbits and Other Wearables May Not Accurately Track Heart Rates in People of Color," STAT, July 24, 2019, https://www.statnews.com/2019/07/24/fitbit-accuracy-dark-skin/.
Peggy Bui and Yuan Liu, "Using AI to Help Find Answers to Common Skin Conditions," Google, May 18, 2021, https://blog.google/technology/health/ai-dermatology-preview-io-2021/.
Yuan Liu et al., "A Deep Learning System for Differential Diagnosis of Skin Diseases," *Nature Medicine* 26, no. 6 (June 2020), https://doi.org/10.1038/s41591-020-0842-3.
237. Todd Feathers, "Google's New Dermatology App Wasn't Designed for People with Darker Skin," Motherboard, May 20, 2021, https://www.vice.com/en/article/m7evmy/googles-new-dermatology-app-wasnt-designed-for-people-with-darker-skin.
238. Pragya Kakani et al., "Allocation of COVID-19 Relief Funding to Disproportionately Black Counties," *Journal of the American Medical Association* 324, no. 10 (September 2020), https://doi.org/10.1001/jama.2020.14978.
Emma Pierson et al., "An Algorithmic Approach to Reducing Unexplained Pain Disparities in Underserved Populations," *Nature Medicine* 27, no. 1 (January 2021), https://doi.org/10.1038/s41591-020-01192-7.
239. Jonathan Cohen et al., "Low LDL Cholesterol in Individuals of African Descent Resulting from Frequent Nonsense Mutations in PCSK9," *Nature Genetics* 37, no. 2 (February 2005), https://doi.org/10.1038/ng1509.
240. "PCSK9-Inhibitor Drug Class that Grew Out of UTSW Research Becomes a Game-Changer for Patient with Extremely High Cholesterol," University of Texas Southwestern Medical Center, February 25, 2016, https://www.utsouthwestern.edu/newsroom/articles/year-2016/pcsk9-patient-khera.html.
241. Layal Liverpool, "Genetic Studies Have Missed Important Gene Variants in African People," *New Scientist*, October 31, 2019, https://www.newscientist.com/article/2221957-genetic-studies-have-missed-important-gene-variants-in-african-people/.
Deepti Gurdasani et al., "Uganda Genome Resource Enables Insights into Population History and Genomic Discovery in Africa," *Cell* 179, no. 4 (October 2019), https://doi.org/10.1016/j.cell.2019.10.004.
242. Melinda C. Mills and Charles Rahal, "A Scientometric Review of Genome-Wide Association Studies," *Communications Biology* 2, no. 1 (2019), https://doi.org/10.1038/s42003-018-0261-x.
Giorgia Sirugo, Scott M. Williams, and Sarah A. Tishkoff, "The Missing Diversity in Human Genetic Studies," *Cell* 177, no. 1 (March 2019), https://doi.org/10.1016/j.cell.2019.04.032.
243. Bill Clinton, "June 2000 White House Event," National Human Genome Research Institute, August 29, 2012, https://www.genome.gov/10001356/june-2000-white-house-event.
Bill Clinton, "Apology For Study Done in Tuskegee," The White House, May 16, 1997, https://clintonwhitehouse4.archives.gov/textonly/New/Remarks/Fri/19970516-898.html.
244. Peter M. Visscher, "10 Years of GWAS Discovery: Biology, Function, and Translation," *American Journal of Human Genetics* 101, no. 1 (July 2017), https://doi.org/10.1016/j.ajhg.2017.06.005.
245. Joannella Morales et al., "A Standardized Framework for Representation of Ancestry Data in Genomics Studies, with Application to the NHGRI-EBI GWAS Catalog," *Genome Biology* 19, no. 1 (February 2018), https://doi.org/10.1186/s13059-018-1396-2.
246. Deborah Borfitz, "H3Africa Consortium Primes a Continent for Large-Scale Genomics Research," H3Africa, February 2, 2021, https://www.bio-itworld.com/news/2021/02/02/h3africa-consortium-primes-a-continent-for-large-scale-genomics-research.
Ananyo Choudhury et al., "High-Depth African Genomes Inform Human Migration and Health," *Nature* 586, no. 7831 (October 2020), https://doi.org/10.1038/s41586-020-2859-7.
247. Aimé Lumaka et al., "Increasing African Genomic Data Generation and Sharing to Resolve Rare and Undiagnosed Diseases in Africa: a Call-To-Action by the H3Africa

Rare Diseases Working Group," *Orphanet Journal of Rare Diseases* 17 (2022), https://doi.org/10.1186/s13023-022-02391-w.

248. Islam Oguz Tuncay et al., "The Genetics of Autism Spectrum Disorder in an East African Familial Cohort," *Cell Genomics* 3, no. 7 (July 2023), https://doi.org/10.1016/j.xgen.2023.100322.

249. Matteo Fumagalli et al., "Greenlandic Inuit Show Genetic Signatures of Diet and Climate Adaptation," *Science* 349, no. 6254 (September 2015), https://doi.org/10.1126/science.aab2319.
PingHsun Hsieh et al., "Adaptive Archaic Introgression of Copy Number Variants and the Discovery of Previously Unknown Human Genes," *Science* 366, no. 6463 (October 2019), https://doi.org/10.1126/science.aax2083.

250. "23andMe Research Innovation Collaborations Program," 23andMe, https://research.23andme.com/research-innovation-collaborations/.
"Novartis and GSK Announce Collaboration to Support Scientific Research into Genetic Diversity in Africa," *Novartis*, 2021, https://www.novartis.com/news/media-releases/novartis-and-gsk-announce-collaboration-support-scientific-research-genetic-diversity-africa.

251. Megan Molteni, "The Massive, Overlooked Potential of African DNA," *Wired*, October 1, 2019, https://www.wired.com/story/the-massive-overlooked-potential-of-african-dna/.

252 Iain Mathieson and Aylwyn Scally, "What Is Ancestry?," *PLoS Genetics* 16, no. 3 (2020), https://doi.org/10.1371/journal.pgen.1008624.
Segun Fatumo et al., "Promoting the Genomic Revolution in Africa through the Nigerian 100K Genome Project," *Nature Genetics* 54, no. 5 (May 2022), https://doi.org/10.1038/s41588-022-01071-6.

253. Jacob Bell, "Drug Companies Pay for Exclusive Access to UK Genetic Data," BioPharma Dive, September 11, 2019, https://www.biopharmadive.com/news/uk-biobank-genome-sequencing-investment-amgen-astrazeneca-johnson-gsk/562709/.
"GSK and 23andMe Sign Agreement to Leverage Genetic Insights for the Development of Novel Medicines," GSK, July 25, 2018, https://www.gsk.com/en-gb/media/press-releases/gsk-and-23andme-sign-agreement-to-leverage-genetic-insights-for-the-development-of-novel-medicines/.
"23andMe Announces Extension of GSK Collaboration and Update on Joint Immuno-oncology Program," 23andMe, 2022, https://investors.23andme.com/news-releases/news-release-details/23andme-announces-extension-gsk-collaboration-and-update-joint.
"Almirall Signs a Strategic Agreement with 23andMe to License Rights of a Bispecific Monoclonal Antibody that Blocks All Three Isoforms of IL-36 Cytokine," Cision PR Newswire, 2020, https://www.prnewswire.com/news-releases/almirall-signs-a-strategic-agreement-with-23andme-to-license-rights-of-a-bispecific-monoclonal-antibody-that-blocks-all-three-isoforms-of-il-36-cytokine-300984521.html.

254. "Significant Research Advances Enabled by HeLa Cells," National Institutes of Health, https://osp.od.nih.gov/hela-cells/significant-research-advances-enabled-by-hela-cells/.
Maninder Ahluwalia, "Genetic Privacy: We Must Learn from the Story of Henrietta Lacks," *New Scientist*, August 1, 2020, https://www.newscientist.com/article/2250449-genetic-privacy-we-must-learn-from-the-story-of-henrietta-lacks/.
Rebecca Skloot, *The Immortal Life of Henrietta Lacks* (New York: Crown, 2010).

255. Rhys Blakely, "Genetics Lab Told to Hand Back African Tribes' DNA," *The Times*, October 14, 2019, https://www.thetimes.co.uk/article/genetics-lab-told-to-hand-back-african-tribes-dna-83xqls5sh.
Erik Stokstad, "Major UK Genetics Lab Accused of Misusing African DNA," *Science*, October 30, 2019, https://www.science.org/content/article/major-uk-genetics-lab-accused-misusing-african-dna.

256. Rebecca Skloot, "The Immortal Life of Henrietta Lacks, the Sequel," *Opinion, New York Times*, March 23, 2013, https://www.nytimes.com/2013/03/24/opinion/sunday/the-immortal-life-of-henrietta-lacks-the-sequel.html.

"Summary of the NIH HeLa Genome Data Use Agreement," National Institutes of Health, 2013, https://www.nih.gov/sites/default/files/institutes/foia/summary-data-use.pdf.

Kathy L. Hudson and Francis S. Collins, "Biospecimen Policy: Family Matters," *Nature* 500, no. 7461 (August 2013), https://doi.org/10.1038/500141a.

257. "Who Are the Uyghurs and Why Is China Being Accused of Genocide?," BBC, May 24, 2022, https://www.bbc.com/news/world-asia-china-22278037.

258. "China: Minority Region Collects DNA from Millions," Human Rights Watch, December 13, 2017, https://www.hrw.org/news/2017/12/13/china-minority-region-collects-dna-millions.

259. Sui-Lee Wee, "China Uses DNA to Track Its People, with the Help of American Expertise," *New York Times*, February 21, 2019, https://www.nytimes.com/2019/02/21/business/china-xinjiang-uighur-dna-thermo-fisher.html.

260. Neil Munshi, "How Unlocking the Secrets of African DNA Could Change the World," *Financial Times*, March 5, 2020, https://www.ft.com/content/eed0555c-5e2b-11ea-b0ab-339c2307bcd4.

"Pfizer, Kano State Reach Settlement of Trovan Cases," Pfizer, 2009, https://www.pfizer.com/news/press-release/press-release-detail/pfizer_kano_state_reach_settlement_of_trovan_cases.

261. "Apology for Study Done in Tuskegee," White House, May 16, 1997, https://clintonwhitehouse4.archives.gov/textonly/New/Remarks/Fri/19970516-898.html.

"The Tuskegee Timeline," Centers for Disease Control and Prevention, 2021, https://www.cdc.gov/tuskegee/timeline.htm.

Jean Heller, "AP Exposes the Tuskegee Syphilis Study: The 50th Anniversary," Associated Press, July 25, 2022, https://apnews.com/article/tuskegee-study-ap-story-investigation-syphilis-53403657e77d76f52df6c2e2892788c9.

"Tuskegee Syphilis Experiment," Equal Justice Initiative, October 31, 2020, https://eji.org/news/history-racial-injustice-tuskegee-syphilis-experiment/.

262. Keolu Fox, "The Illusion of Inclusion—the 'All of Us' Research Program and Indigenous Peoples' DNA," *New England Journal of Medicine* 383, no. 5 (July 2020), https://doi.org/10.1056/NEJMp1915987.

263. Kelly D. Myers et al., "Effect of Access to Prescribed PCSK9 Inhibitors on Cardiovascular Outcomes," *Circulation: Cardiovascular Quality and Outcomes* 12, no. 8 (August 2019), https://doi.org/10.1161/circoutcomes.118.005404.

264. "Native BioData Consortium Research," Native BioData Consortium, 2021, https://nativebio.org/research/.

Christine A. Peschken et al., "Rheumatoid Arthritis in a North American Native Population: Longitudinal Followup and Comparison with a White Population," *Journal of Rheumatology* 37, no. 8 (August 2010), https://doi.org/10.3899/jrheum.091452.

265. Rex Dalton, "Tribe Blasts 'Exploitation' of Blood Samples," *Nature* 420, no. 6912 (November 2002), https://doi.org/10.1038/420111a.

Edward F. Foulks, "Misalliances in the Barrow Alcohol Study," *American Indian and Alaska Native Mental Health Research* 2, no. 3 (1989), https://doi.org/10.5820/aian.0203.1989.7.

266. Zachary Tracer and Erin Brodwin, "A Tiny Startup Wants to Pay You for Your DNA, and It Could Lead to the Next Wave of Medical Innovation," *Business Insider*, December 15, 2018, https://www.businessinsider.com/lunadna-pays-for-gene.

267. Serena Williams, "How Serena Williams Saved Her Own Life," *Elle*, April 5, 2022, https://www.elle.com/life-love/a39586444/how-serena-williams-saved-her-own-life/.

268. "Reproductive Injustice: Racial and Gender Discrimination in US Health Care," Center for Reproductive Rights, 2014, https://reproductiverights.org/wp-content/uploads/2020/12/CERD_Shadow_US_6.30.14_Web.pdf.

269. Evelyn F. Forget, "The Town with No Poverty: The Health Effects of a Canadian Guaranteed Annual Income Field Experiment," *Canadian Public Policy* 37, no. 3 (September 2011), https://doi.org/10.3138/cpp.37.3.283.

Arne Ruckert, Chau Huynh, and Ronald Labonté, "Reducing Health Inequities: Is Universal Basic Income the Way Forward?," *Journal of Public Health* 40, no. 1 (March 2018), https://doi.org/10.1093/pubmed/fdx006.

Marcia Gibson, Wendy Heart, and Peter Craig, "Potential Effects of Universal Basic Income: A Scoping Review of Evidence on Impacts and Study Characteristics," *Lancet* 392 (November 2018), https://doi.org/10.1016/S0140-6736(18)32083-X.

Olli Kangas et al., "The Basic Income Experiment 2017–2018 in Finland: Preliminary Results," Finnish Ministry of Social Affairs and Health, 2019, https://ubiru.org/wp-content/uploads/2020/03/2019_Finland_Report_The-Basic-Income-Experiment-2017-2018-in-Finland.pdf.

Darrick Hamilton and William Darity Jr., "Can 'Baby Bonds' Eliminate the Racial Wealth Gap in Putative Post-Racial America?," *Review of Black Political Economy* 37, no. 3 (September 2010), https://doi.org/10.1007/s12114-010-9063-1.

Shira Markoff et al., "A Brighter Future with Baby Bonds: How States and Cities Should Invest in Our Kids," 2022, https://prosperitynow.org/sites/default/files/resources/A-Brighter-Future-With-Baby-Bonds_2.pdf.

270. Kenya Evelyn, "'Like I Wasn't There': Climate Activist Vanessa Nakate on Being Erased from a Movement," *Guardian*, January 29, 2020, https://www.theguardian.com/world/2020/jan/29/vanessa-nakate-interview-climate-activism-cropped-photo-davos.

271. "Leading MPs sign Early Day Motion in support of Ella's Law," *Ella's Law*, December 13, 2022, https://ellaslaw.uk/2022/12/13/leading-mps-sign-early-day-motion-in-support-of-ellas-law/.

272. Xavier Lopez, "How a Medical Student Is Using TikTok to Bridge Racial Health Disparities," *Pulse*, June 17, 2022, https://whyy.org/segments/how-a-medical-student-is-using-tiktok-to-bridge-racial-health-disparities/.

About the Author

LAYAL LIVERPOOL is a science journalist with expertise in biomedical science, particularly virology and immunology. Her PhD research at Oxford focused on investigating how invading viruses are detected by the body's immune system. Her writing has appeared in *Nature*, *New Scientist*, *Wired*, and *The Guardian*.